But *Why?*

But *Why?*

UNRAVELING

the Mysteries

of

Mathematics

Sean Monroe

Copyright © 2011 by Sean Monroe.

Library of Congress Control Number: 2011914421
ISBN: Hardcover   978-1-4653-5043-5
              Softcover   978-1-4653-5042-8
              Ebook   978-1-4653-5044-2

All rights reserved. No part of this book may be reproduced or transmitted in any form or by any means, electronic or mechanical, including photocopying, recording, or by any information storage and retrieval system, without permission in writing from the copyright owner.

This book was printed in the United States of America.

Sean Monroe would love to hear from you. You may contact him at mrmathmonroe@gmail.com.

To order additional copies of this book, contact:
Xlibris Corporation
1-888-795-4274
www.Xlibris.com
Orders@Xlibris.com

To all those who have endured discussions about mathematics with me;
To my friends in the mathematical world for your acceptance;
To my teachers for your dedication;
To those who love mathematics and helped me love it more;
To those who have difficulty with mathematics who gave me focus;
To all those who spent countless hours editing my book (and/or using it as a non-addictive sleeping aid);
To my family for being my inspiration;

I thank you and dedicate this book to you.

Introduction

Why the "Why"?

A few years ago, I was in a high school classroom where the teacher was explaining how to divide with decimals. One of the students asked the question, "Why do we do that—move the decimal point in the divisor?" While the question intrigued me, the teacher's response concerned me: "That is just the way we do it."

Later that year, I was visiting another teacher, and a student asked, "Why do we count decimal places when multiplying with decimals?" Again the teacher gave a stoic response, "That is how I learned it."

At first I was just concerned for the lack of explanation, but then I came to realize that most teachers, including myself, did not know some of the reasons why. Consequently, I found myself on a quest to find simple explanations for some of the "why" questions that students might bring up.

Additionally, I found that even if teachers knew the reasons, they did not expound on them in class. They would give a reason similar to the ones shared above and then move on, sacrificing the explanation for the time needed to get through the curriculum.

To address this issue, I started to make a list of what came to be known as the mysteries in the mathematics classroom. They were called mysteries because they were treated as such by the teachers in the classroom. With no formal explanation, these concepts would become just that, a mystery, to anyone who either asked the question why, thought to ask why, or didn't know to ask why.

There are undoubtedly many ways to explain some of these mysteries to students (and probably more efficient than the ones in this book) though during my research, some seemed to be a bit elusive. I collaborated with many of my colleagues to explore different points of view of some of the more trivial concepts in order to help me see them in a way that would birth a coherent and concise explanation.

I also want to make it clear that this book is not a book that proves everything. I intentionally steered away from using formal proof in this book because I wanted to make the resource accessible to as many people as possible. In fact, many of the mysteries explored in this resource are not provable. This means that they are accepted in the mathematical world as conventions and have been set forth as "just the way we do it" kinds of things. This resource intends to help give meaning to some of these as well with the intent of removing as much of the mystery in mathematics as possible.

It is Worth the Time

It was the first day of summer school, and if you know the clientele, it is not hard to imagine a class full of disinterested students. I decided to ask some of the students some "*why*" questions. The one that got their attention was, "Why is a negative number times a positive number equal to a negative number?" The moment I asked that question, many heads popped up. "Yeah, I asked my teacher that last year, and she just said, '*Because.*' I stopped listening to her after that." Or "Who comes up with these rules anyway?"

Many more questions were shouted out, and I knew I had baited the hook. I proceeded to discuss with them both the short answer and the detailed answer. Most of them had smiles on their faces and then said, "Teach us something else, like . . . " Even those students that did not fully understand the discussion said things like, "No one has ever tried to explain those kinds of things to us before. Thank you."

The Long and the Short of It

For some of the explanations, there will be a short answer and a more detailed answer. There are several benefits for having both a long and short answer. A teacher may find that the detailed answer is too long to give to students and that the short answer will suffice. So the short answer is provided to assist in explaining the history and development of the concept without delving into a long drawn-out explanation.

The detailed answer should be used to gain more insight or to view optional pedagogical attempts to explain a certain concept. In any case, please enjoy the book and don't be afraid to ask, "*Why?*"

Glossary

At the end of the book, you will find a very short glossary. Originally, I included definitions within the text itself, but I found that many explanations flowed more easily without continually having to explain words.

Questions to Ponder

For each mystery, I have included a set of questions for you to ponder over that will assist in facilitating reflection and discussion. The idea is to give teachers entry points to discussion and to assist those that might use this resource in workshops or professional development.

What is a "Mystery" in Mathematics?

Simple Answer

Even though many will say that everything in math is a mystery, I would like to formalize the definition here so everyone is on the same page: mysteries in mathematics are concepts, procedures and steps, techniques, conventions, and vocabularies used in mathematics that commonly are not thoroughly explained. Meaning, teachers explain *how* to do something, but not *why* we do it; they explain *what* something is, but not *how* we know it; they explain *what* something is called, but not *why* we call it that; and so on.

Detailed Answer

The following vignette will illustrate when a mystery of mathematics might come up in the classroom setting. We join the lesson already in progress.

Let's say we have the following problem:

$$\frac{1}{3} \div \frac{2}{5}$$

Here are the steps on how to divide with fractions:

First, change the division sign to multiplication (because we don't divide with fractions!), and second, flip the second fraction like so:

$$\frac{1}{3} \div \frac{2}{5} \rightarrow \frac{1}{3} \times \frac{5}{2}$$

Third, multiply the numerators and place that number in the numerator of the answer and then multiply the denominators and place that number in the denominator of the answer, like this:

$$\frac{1}{3} \times \frac{5}{2} = \frac{1 \times 5}{3 \times 2} = \frac{5}{6}$$

The teacher will then have the students repeat the steps over and over until they can mimic the steps on their own.

It would not be completely uncommon for a student, during some point of the lesson, to ask a question like this: "Why do we switch the division sign to multiplication?" or "Why do we flip the second fraction?" Many times, the answers to questions such as this, from the teacher's mouth, are the following:

- "That is just the way we do it."
- "Just because."
- "That's the rule."
- "I don't know. That is how I have always done it."
- "That is how I was taught it."

To many students, an answer like this simply adds to the reputation of frustration and confusion that mathematics has been labeled with for so long.

What to Do When a Student Asks, "Why?"

When a student asks a teacher to give a reason for a procedure or a step, the teacher should have a way to cope with that question. There are several ways to handle this situation, You could (a) tell the student one of the reasons above, (b) tell the student that you will find out and come back later with an answer, or (c) help the student figure out the reason.

This book attempts to assist teachers and students with options b and c by first looking closer at some of the possible questions that could come up in a classroom setting and then developing some reasoning behind them. But the overall purpose of this book is to help encourage all involved (teachers, students, and parents) to never take for granted what we do in mathematics. Don't just accept what is being shown to you; ask, "Why?"

Objections
We Don't Have Time!

Anytime students are presented concepts with a little bit of discovery behind it, there are always going to be teachers who question its usefulness because of time constraints. There are a lot of standards to cover, but there is a finite amount of time. This is a very common argument. This is also a very old argument. While I agree that there are some perplexities that can be superficially covered because of time constraints, we can't use this reason for all of them. Teaching a student who is having trouble remembering how to write his numbers a simple explanation or discussion on the reason behind it can help the student better remember and might even make mathematics all the more intriguing.

Why Should I Teach It This Way if Most of My Students Can Memorize Them with All the Activities that I Do?

Besides the reason given above about the few students who struggle to memorize, there is one more reason for doing this: it encourages students to be inquisitive. Sowing an inquisitive attitude into the different components of mathematics will reap the benefit of students who are curious and love to learn. This inquisitiveness is necessary to understanding the more complex concepts in mathematics.

I know by experience that all students can handle the explanations. I know by experience that those students who have been shown some sort of explanation retain the procedure, the steps, and the concept at a higher rate. Even if the students don't remember the reason behind a concept, they will be left with a sense that there is a reason and not be left with the feeling that mathematics is mysterious.

Math is Not Mysterious

Finally, a teacher risks introducing mathematics as a subject based on a mysterious foundation if he or she simply excuses away an explanation with a response such as "That's just the way we do it" or "I don't know" and then doesn't go and find out the reason and share it later. Mathematics is the polar opposite of mysterious; everything in mathematics has a reason.

A Stepping Stone to the Common Core

In June of 2010, the Common Core State Standards were finalized and presented to the states for possible adoption. Since then, several states in the nation have officially adopted them. These standards were written for both mathematics and English language arts. For mathematics, the standards are broken up into two parts: the content portion and the mathematical practices portion. I would claim that this book is a perfect companion to all those that look to teach or learn from the new Common Core State Standards.

The STANDARDS FOR MATHEMATICAL PRACTICE "describe varieties of expertise that mathematics educators at all levels should seek to develop in their students." There are eight STANDARDS FOR MATHEMATICAL PRACTICE, and they are as follows:

1. Make sense of problems and persevere in solving them.
2. Reason abstractly and quantitatively.
3. Construct viable arguments and critique the reasoning of others.
4. Model with mathematics.
5. Use appropriate tools strategically.
6. Attend to precision.
7. Look for and make use of structure.
8. Look for and express regularity in repeated reasoning.

But Why? Unraveling the Mysteries of Mathematics will help develop these STANDARDS FOR MATHEMATICAL PRACTICE in the students by assisting the teachers with natural entry points to discussion, pattern finding, repeated reasoning, constructing arguments, reasoning abstractly, persevering in problem solving, and using appropriate tools to understand mathematics.

Many of the mysteries are addressed directly in the content portion of the standards as well. For example, there is a standard in fifth grade that calls for students to "add, subtract, multiply, and divide decimals . . . using concrete models or drawings and strategies based on place value"; and the eleventh, twelfth, and thirteenth mysteries explicitly address this standard in a way that is consistent with the way the Common Core asks that these concepts be addressed.

In no way should this book take the place of any textbook or curriculum that is state adopted for use with the Common Core State Standards. This book could be used as a supplement to the concepts taught in those resources and be a resource for teachers, parents, and students alike.

The Mysteries of Mathematics

I have endeavored to put these mysteries in a logical order. Many of the concepts should sound familiar. If not, have fun with them anyway.

MYSTERY #1
What Is and Why Do We Call It *Mathematics?* 21

MYSTERY #2
What Is and Why Do We Call It *Algebra?* 25

MYSTERY #3
What Is and Why Do We Call It *Geometry?* 29

MYSTERY #4
What Is and Why Do We Call It *Trigonometry?* 31

MYSTERY #5
What Is and Why Do We Call It *Calculus?* 33

MYSTERY #6
Why Do We Write the Numbers 0–9 That Way? 35

MYSTERY #7
What Does It Mean to Regroup, Borrow, or Carry? 40

MYSTERY #8
Why Can't We Divide by Zero? 46

MYSTERY #9
    Why Do We Do It That Way?
        Long Division.. 50

MYSTERY #10
    Why Do We Do It That Way?
    "Invert and Multiply"
    when Dividing by a Fraction...................................... 58

MYSTERY #11
    Why Do We Do It That Way?
    Converting a Mixed Number to an
    Improper Fraction 61

MYSTERY #12
    Why Do We Do It That Way?
    Converting an Improper Fraction
    to a Mixed Number... 65

MYSTERY #13
    Why Do We Do It That Way?
    Counting Decimal Places when
    Multiplying with Decimals 68

MYSTERY #14
    Why Do We Do It That Way?
    Moving the Decimal Point in the Divisor
    when Dividing by a Decimal...................................... 71

MYSTERY #15
    Why Is 0 Times Any Number Equal to 0? 74

MYSTERY #16
    Why Is the Product of a Positive Integer and a
        Negative Integer Equal to a Negative Integer?....... 76

MYSTERY #17
    Why Is the Product of a Negative Integer and a
        Negative Integer Equal to a Positive Integer?........ 78

MYSTERY #18
        Why Is Any Number Raised to the Zero Power
            Equal to One, or in Other Words,
            Why Is $n^0 = 1$? 80

MYSTERY #19
        Why Is Any Number Raised to a Negative One
            Equal to Its Reciprocal, or in Other Words,
            Why Is $n^{-1} = \frac{1}{n}$? 84

MYSTERY #20
        Why Do We Do It That Way?
            Make two equations when Solving Equations
            or Inequalities with Absolute Values 88

MYSTERY #21
        What Is PEMDAS? 91

MYSTERY #22
        What Is FOIL? 95

MYSTERY #23
        What Does It Mean to "Complete the Square"? 99

MYSTERY #24
        Why Does $0! = 1$? 103

MYSTERY #25
        What Is π? 107

MYSTERY #26
        What Is e? 110

MYSTERY #27
        What Is i? 113

MYSTERY #28
        Why Are There 360° in a Circle? 116

MYSTERY #29
> What Is a Radian? 118

MYSTERY #30
> Why Are There 2π Radians in a Circle? 121

Now we will delve into some of these mysteries of mathematics. We will use many different methods to help make sense of these rarely explained concepts in mathematics. We will use patterns, manipulatives, graphs, diagrams, and sample questions to illustrate the reasons behind these ideas. These will not be all of the ways to explain the mysteries, but it will be a good start. Let's begin.

Mystery #1

What Is and Why Do We Call It Mathematics?

Why Bother?

Really, why bother with discussing the reasons why we call this subject mathematics? And this is where we will begin. We will begin with this seemingly unimportant question to help focus the book. We need to know the reason why we call this subject mathematics. We need to know what this subject is really all about. Students seem to always ask, "Why are we learning this?" or "When will we ever use this?" The definition of the word *mathematics* may assist us in answering those questions as well.

Short Answer

The word *mathematics* comes from the Greek word *máthēma*, which means learning, study, and science or from the Latin word *mathmaticus*, which means mathematical. Since much of our knowledge of mathematics has originated from the Greek and other ancient civilizations, it is easy to accept the fact that we would use the word that they used for the subject. The fact that mathematics means learning helps us understand one of the reasons teachers give to those skeptical of mathematics for learning mathematics: it teaches us to think. Mathematics, more than any other subject, assists in helping pupils learn, think logically, be forward minded, and communicate precisely. All of which is very much a part of learning.

Detailed Answer

Besides what is mentioned above, it is beneficial to discuss what mathematics is and isn't, especially in the context of this book. First of all, many people see mathematics today in two different phases: elementary mathematics and high school mathematics. We will discuss a third (college mathematics) in a moment. Possibly, a better way to describe mathematics, if we were to attempt some sort of a division of the topic, is to simply state that mathematics is both arithmetic and logic. I will explain.

Math Is Arithmetic

This is where we learn to add, subtract, multiply, and divide. Additionally, we first learn to write numbers and to count and even use numbers to describe the world (usually in the form of word problems). Most of this is what we call arithmetic. Arithmetic is the computational part of mathematics. Whenever we add, subtract, multiply, or divide (or any variation of these four basic operations), we are doing arithmetic. In general, the mathematics learned in elementary school *is* arithmetic. It is true that students will learn more than just arithmetic during their elementary years, but for the most part, the purpose of elementary mathematics is to teach arithmetic.

Math Is Logic

Dividing mathematics up into age levels erroneously suggests that there is a very clean division between the time we learn arithmetic and the time we learn logic. This suggests that the learning of arithmetic has an end and then there is a definite beginning to learning logic. In the context of this book, logic is that which applies the arithmetic to solve a problem, find a pattern, or assist in discovering an idea. Students in elementary school are asked to think logically when they encounter problems that require deciphering *which* operation to use to solve a problem. Arithmetic by itself means very little, but together with logic, we get mathematics.

To put it frankly, simply learning arithmetic is not enough. I equate it to learning the alphabet and spelling words without being able to read and comprehend what you are reading or without learning to write a coherent sentence. If we simply learn arithmetic and don't learn the logic, we are not learning mathematics and, thus, cannot take advantage of the many uses of mathematics.

One of the big questions is, Which should come first—arithmetic or logic? Some may not see an easy black-or-white answer to this question. Many curricula today have them taught together. For example, as students are taught arithmetic, they can learn how to decipher when to use an operation, and as students are taught problem solving, they can be taught arithmetic.

So onto high school mathematics: it is usually divided up into courses. These courses carry titles such as algebra, geometry, precalculus, calculus, trigonometry, and so on. Some of these will be formally defined later, but simply put, they are subsets of one thing: mathematics. The high school courses listed above make up the "logic" portion of mathematics. In all those courses, we apply the arithmetic taught in elementary school. This does not take into account the many topics discussed in university- and college-level courses.

As stated before, dividing mathematics into two groups—elementary (or arithmetic) and high school (or logic)—forces a clear break between the two, one that simply does not exist. Arithmetic and logic go hand in hand. They cannot live without each other. They must remain together. Keep the relationship between arithmetic and logic in mind as we discuss some of the mysteries in this book and use one to discuss the other.

Any definition of mathematics would not be complete without the following statement: math is a science and a language.

Math Is a Science

How is it a science? Observations, experiments, discoveries, and conjectures are as much a part of mathematics as any other natural science. There are also a lot of trial and error, hypothesis and investigation, measurement, classification, and verification. Roger Bacon once said that "mathematics is the gate and key of the sciences . . . [and without it we] cannot know the other sciences or the things of this world."

It is a science of patterns. Mathematics anticipates nature. Mathematics ensures that any pattern encountered by scientists will be explained somewhere as part of the practice of mathematics.

Math Is a Language

Nature speaks to us through mathematics. Mathematics describes the patterns that occur in the natural world, helping us understand how the world works.

Mathematics is also the language of the business world. Economists, businesses, and the financial world in general cannot predict sales, revenue, attendance, or profit without mathematics. Every time a purchase is made at a store, online, or over the phone, mathematics is working hard to make our world work.

Math Is Pattern-Finding

Ultimately, mathematics is the search for patterns. Mathematicians are pattern-finders. One could say with confidence that almost everything we do in mathematics is the result of someone recognizing a pattern and applying the language of mathematics to it. Throughout this book, keep this in mind, for we will use patterns to unravel the underlying mathematics hidden in much that we do in mathematics.

Questions to Ponder

1. Why do we call it mathematics?
2. If someone asked you to define mathematics, how would you answer?
3. What is your view of mathematics? (i.e., How do you feel about mathematics?)
4. What are some of the experiences (good and bad) that you have had with mathematics?
5. When do you use math in your life (e.g., at work, at school, at home)?

Mystery #2

What Is and Why Do We Call It Algebra?

Why Bother?

Again, does it really matter what we call this subject? If we called it difficult or frustrating, the reader might more readily agree with the name of the subject. However, knowing the origins of the actual name of the topic might provide some insight on what algebra is. Many see it as just a bunch of letters, equations, and formulas and a topic that is just plain hard.

Simple Answer

The word *algebra* comes from a book written by mathematician Muhammad ibn Musa al-Khowarizmi. The book is entitled *Hidab al-jabr wal-muqubala* or *The Book of Restoration and Balancing*. The word we use today for this course of study is derived from the word *al-jabr* in the title.

In essence, his book was the first algebra book because it was the first that described the simple techniques that we still use today for solving equations.

Detailed Answer

Mathematician Muhammad ibn Musa al-Khowarizmi wrote the book *Hidab al-jabr wal-muqubala* in AD 825. The title of the book is translated as *The Book of Restoration and Balancing*.

The words *al-jabr* and *muqubalah* were used by the Arab mathematician al-Khowarizmi to describe two operations in manipulating equations in an effort to solve them. *al-Jabr* was the word to describe the action of transposing subtracted terms to the opposite side of the equation. *Muqubalah* was the term used to describe the action of canceling like terms on opposite sides of the equation.

As equations were solved using these two operations, eventually one of the names stuck, *al-jabr*, while the others fell away.

Let me show you an example using the two words:

$$3x + 3 = 9$$
$$\underline{-3 \quad -3} \quad \longleftarrow \text{This is the *al-jabr* step.}$$
$$3x = 9 - 3 \quad \longleftarrow \text{This is the *muqubalah* step.}$$
$$3x = 6$$
$$x = 2$$

We don't use these terms today, but the techniques are still in use.

Interesting Note

The word *algorithm*—the word most commonly used for describing a set of procedures for finding a value or calculating a function—may derive from the name of the author of the book mentioned above, *al-Khowarizmi*.

It may have also come from a treatise that *al-Khowarizmi* wrote in AD 825 entitled *On Calculation with Hindu Numerals*, which translated into Latin is *Algoritmi de numero Indorum*.

But What Is Algebra?

Before we define algebra, it is probably necessary to address the fact that it is generally viewed with detestation. For example, an editor once wrote the following when asked to describe algebra:

> If there is a heaven for school subjects, algebra will never go there. It is the one subject in the curriculum that has kept children

from finishing high school, from developing their special interests, and from enjoying much of their home study work. It has caused more family rows, more tears, more heartaches, and more sleepless nights than any other school subject. (An anonymous editorial writer, ca. 1936, Greenes, page 3)

Algebra has been described as generalized arithmetic, the process for which we solve polynomial equations, and a language for describing properties of functions and graphs. Whichever the definition, the heart of the heartache is the abstractness for which the subject is presented. Algebra is most often presented via symbolization and rigid properties that for many people are hard to "see." *Abstractness* is defined as the process of expressing ideas intrinsically with no attempt at representing it pictorially or in any other visual form. Let me explain.

If I write down the following summation, $1 + 2 = 3$, I have done so with the abstract symbols: 1, 2, +, and 3. These symbols have meaning only because we have defined them with such. The 1 represents in most cases one object. The 2 represents two objects, and the 3 represents three objects. The + sign represents an operation that we define as addition or putting things together. Students are usually not introduced to these symbols without physical objects to help them grasp their meanings. A teacher may say "one" and hold up an object and then write down the symbol 1 on the board and say "one" again. Eventually, students learn to associate the concept of one object with its symbol, 1.

Algebra is generally presented abstractly. If students have a hard time transferring the symbolic representation to the tangible and vice versa, this can cause frustration for the students. Students who struggle solving equations may have issues with the abstract part of the process and may be assisted with some tangible objects, such as algebra tiles.

Algebra as Generalized Arithmetic

Algebra can be characterized as symbolic representation of the four basic operations of addition, subtraction, multiplication, and division. This definition of algebra is the process for which we generalize operations with letters or other symbols without being constrained to a specific situation. For example, in the problem we presented above, $1 + 2 = 3$, we have a specific situation. But we can represent this situation with letters (which we call variables) in this manner: $a + b = c$. The summation equation $a + b = c$ can be read as "The sum of two values will yield a third value." Generalized arithmetic allows for variations in the situation such as $b = c - a$ or $a = c - b$.

Algebra as the Process for Solving Polynomial Equations

"Solving for x" seems to be what most people remember as algebra. An equation is given, and the goal of the problem is to find out what that elusive x represents. This is the essence of algebra. Most of what we learn in solving equations is used throughout all algebra courses. However, algebra is not just solving equations as we have already mentioned, but it is a good portion of it.

Algebra as the Language for Describing the Properties of Functions and Graphs

Algebra is a language. It is a language for generalizing arithmetic, for solving equations, and for describing the properties of functions and graphs. It was never my intent to define functions or graphs at this point but only to define algebra as its language. However, it may be necessary to at least define functions and graphs as a way of describing relationships between two or more things. Functions and graphs use the language of algebra to define these relationships.

Questions to Ponder

1. If someone asked you to define algebra, how would you answer?
2. Why do we call it algebra?
3. What was your experience with algebra?
4. What concepts do you remember from algebra? Why do you think you remember them?
5. Which topic was your favorite? Why?
6. Which topic was your least favorite? Why?
7. What first comes to mind when you think of algebra?

Mystery #3

What Is and Why Do We Call It Geometry?

Why Bother?

Years ago there were commercials that asked the question, "Have you ever wondered?" These short clips answered questions like Where do crayons come from? or Why do we hiccup? Whether you have wondered about the meaning of the word *geometry* or not, knowing the actual meaning of the word will give insight into what the topic really is all about and provide a clearer understanding of why we study it.

Short Answer

The word *geometry* can be broken up into Greek roots for "earth" and "to measure" that come from the word *geometria—gë* and *metreein* respectively. Therefore, the word *geometry* means the study of measuring the earth. This, when one thinks about it, is the perfect name for a subject that requires students to describe, measure, and analyze objects, both two-dimensionally and three-dimensionally.

Detailed Answer

Research suggests that the foundations of geometry started in the Mesopotamia region around 3500 BCE. In fact, the Mesopotamians were among the earliest peoples to know about what is called the Pythagorean theorem.

Most of what we attribute to geometry, however, comes from a book called *The Elements*, written by the mathematician Euclid. *The Elements* is a collection of thirteen books on geometry and other mathematics. The first six books offer explanations and proofs of the foundations of elementary plane geometry with sections on triangles, rectangles, circles, polygons, proportions, congruence, and similarities. The rest of the books present other mathematical ideas, including solid geometry, pyramids, and Platonic solids. The common high school geometry book used today is based on this book.

I often wonder why we didn't call this course of study elements after Euclid's book, but geometry seems to fit better anyway.

Questions to Ponder

1. Why do we call it geometry?
2. If someone asked you to define geometry, how would you answer?
3. What was your experience like with geometry?
4. How would you compare algebra and geometry?
5. Which concept was your favorite in geometry? Why?
6. Which concept was your least favorite in geometry? Why?

Mystery #4

What Is and Why Do We Call It Trigonometry?

Why Bother?

What's in a name? In the case of trigonometry, there is quite a bit. We bother with defining trigonometry and discussing the origins of its name because it gives us a much deeper understanding of what the teacher is talking about during the class. So if you continue, we will discuss the history of the word and how it relates to topics discussed in the course.

An Answer

In short, trigonometry is the study of triangles. The word *trigonometry* can be broken down into its Greek roots: *tri* means three, *gon* means angles, and *metro* means measure. So the word *trigonometry* means the measure of triangles.

Trigonometry is the study of how the sides and angles of a triangle are related to one another. It has numerous applications in fields such as astronomy, maritime, surveying, aerial navigation, and engineering.

Trigonometry is also the study of "special functions' called trigonometric functions such as sine, cosine, and tangent.

Questions to Ponder

1. If someone asked you to define trigonometry, how would you answer?
2. Do you remember anything from trigonometry? Why or why not?
3. Do you think trigonometry is important? Why or why not?
4. Why do we call it trigonometry?
5. What would the study of quadrilaterals be called? What would the study of pentagons be called?

MYSTERY #5

What Is and Why Do We Call It Calculus?

Where Is the Mystery?

The mystery here lies in any potential benefit in knowing the meaning of the word *calculus*. Many times I have found that learning the basic meaning of a word has deepened my understanding of not only the word but also the context in which it is used. This is definitely the case with the word *calculus*. So what is the meaning of the word *calculus*?

An Answer

In Latin, calculus is defined as a small stone used for counting. Today, calculus is one of the courses, usually one of the last, taken in high school and in the first years of college.

In a way, calculus is the perfect name for the course also known as infinitesimal analysis. Calculus uses the concept of infinitesimals to do its calculations. Calculus uses this concept to calculate effectively the area and volume of shapes and calculate the speed and acceleration of objects. It accomplishes this by essentially counting up very small rectangles used to form that object. For example one can find the area under the curve below by drawing rectangles as shown.

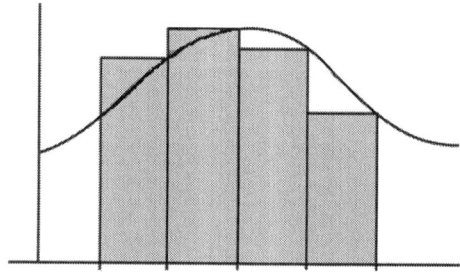

The rectangles above only give an estimate of the area under the curve shown. Calculus uses techniques that draws infinitesimal sized rectangles under the curve and then sums up the rectangles to arrive at the area under the curve. Those little rectangles, or very small "stones", give Calculus its name.

Questions to Ponder

1. If someone asked you to define calculus, how would you answer?
2. What first comes to mind when you think of calculus?
3. If you took calculus, what do you remember? Why do you think you remember it?
4. Do you think calculus is a good name for this subject? Why or why not?
5. Do you think that students have a negative view of calculus?
6. What do you think students think of calculus?

MYSTERY #6

Why Do We Write the Numbers 0-9 That Way?

Where Is the Mystery?

The mystery is the very fact that all we do is tell kids how to write the numbers. The mystery is the fact that teachers, including myself for a time, usually don't know the origins of the symbols we use. And this is where the mysteries of mathematics begin. We simply tell the students how to write the numbers without involving them in their development. Sure, the students will go on just fine without knowing any reason for their shape or size, but the seed of mystery is planted with such a seemingly innocent tactic. Many great mathematicians do not know their origins, but what an amazing discussion a teacher can have with their students if you just allow them to ask: "Why do we write the number 1 that way?" Or just ask it yourself.

Short Answer

There are many ways of writing each number. The current number system is based on the Hindu-Arabic system, which became popular when a mathematician named Fibonacci saw them used in Africa. It is suggested that this system may have derived from the number of angles used to form the numbers.

Detailed Answer

The first thing we must recognize is that there is more than one way to write each number. Even though many call mathematics the universal language, there are many different ways of expressing the ideas and concepts in mathematics. The manner for which we write our numbers, for example, is only one way among many current ways and is quite different from how it was done in the past. We will not go into the entire history (books have been written on the subject), but we will look at the orthography of two number systems, which have influenced our own: roman and Hindu-Arabic.

The Roman Numerals

The Romans ruled the known world for many years and have influenced modern life more than almost any other ancient civilization. The number system used in Rome and throughout its empire is still used today in certain settings, such as on watches and clocks, outlines, and big events. A table can be found below.

This form of writing the numbers was very efficient for representing numbers, but it had its limitations. For example, when it came to computing, roman numerals can become very cumbersome. For this reason, some speculate that the roman numerals were only used to represent the numbers and not used for computation at all. A more convenient way of writing numbers was needed; enter India.

Roman Numeral	Common Way
I	1
II	2
III	3
IV	4
V	5
VI	6
VII	7
VIII	8
IX	9
X	10
L	50
C	100
D	500
M	1,000

Hindu-Arabic Numerals

The form that we use today was developed over time in the Arab and Indian societies. It was first introduced in Europe by the Arabs, and so they are sometimes commonly referred to as the *arabic numerals*. Actually, it was the famous Italian mathematician Fibonacci who introduced Europe to this type of writing when he claimed their usefulness in his book *Liber Abaci*, released in 1202. Each number went through a series of changes before they arrived at their present appearance. Many tables and illustrations exist that will show you the evolution of their appearance. Below is one such table.

$$\begin{array}{ccccccccc} I & 乙 & \xi & \mathcal{V} & \mathsf{U} & b & 7 & 8 & 9 \\ \prime & 乙 & 3 & \mathsf{F} & 9 & \mathsf{L} & \wedge & 8 & 2 \\ 1 & 乙 & 3 & \mathsf{lf} & 4 & \mathsf{P} & \mathsf{r} & 8 & 9 \\ 1 & 7 & 3 & \mathcal{R} & 4 & 6 & \wedge & 8 & 9 \\ 1 & Z & 3 & \mathcal{R} & 4 & 6 & \wedge & 8 & 9 \\ 1 & 2 & 3 & 4 & 4 & 6 & 7 & 8 & 9 \\ 1 & 2 & 3 & 4 & 5 & 6 & 7 & 8 & 9 \end{array}$$

The above table shows the many different forms that existed in the Arabic and Indian societies throughout the years with our current form at the bottom.

Classroom Application

One of the first things that students learn in mathematics is how to count and write the numbers 1-10. A question that may come up as a teacher is helping young uncoordinated hands write the numbers is, Why do we write the numbers that way? In fact, a rich conversation with this age group could come out of simply asking the students the question, "Why do you think we write the one that way?" or "Why do you think we write the two that way?" and so on.

The history given above as to the way we write the numbers today is much more sophisticated than a teacher would probably want to explore with pre-K through kindergarten-aged students. However, an alternative approach might at least give the students an understanding and make some meaning for the students as to why we write the numbers the way we do.

The following is an explanation based on research but not entirely provable that can be used as a justification for the numbers being written the way they are: 0, 1, 2, 3, 4, 5, 6, 7, 8, and 9. The dots are placed to denote an angle.

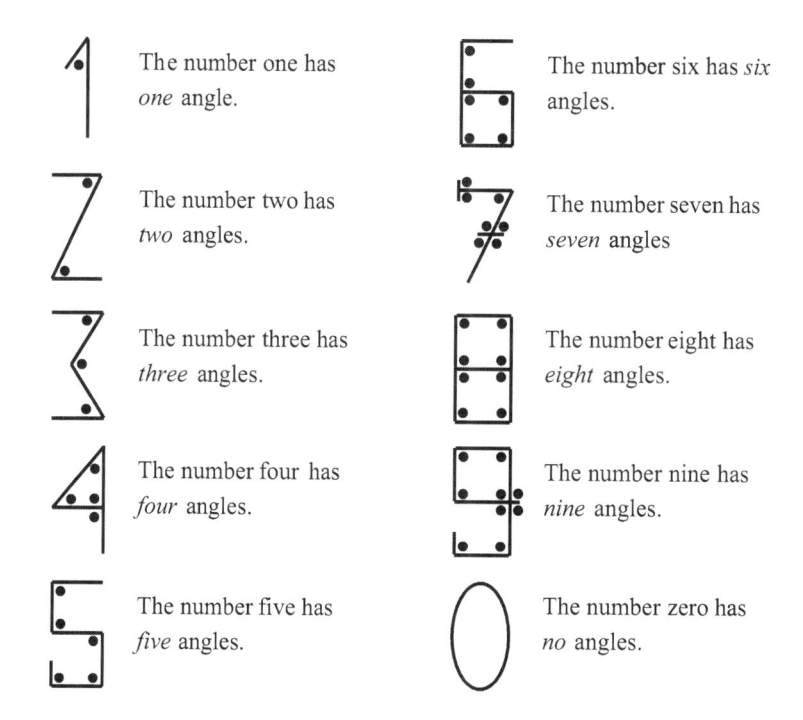

The number one has *one* angle.	The number six has *six* angles.
The number two has *two* angles.	The number seven has *seven* angles
The number three has *three* angles.	The number eight has *eight* angles.
The number four has *four* angles.	The number nine has *nine* angles.
The number five has *five* angles.	The number zero has *no* angles.

This method uses the concept of angles. With kindergarten-aged children, a teacher may elect to simply call angles "corners" and then eventually introduce the students to their proper name, angles.

Interesting Note

If you look back at the chart of Hindu-Arabic numerals, you will notice that zero is not present. Zero is a relatively new "discovery" in the world of mathematics, at least in relation to the other numbers 1–9. However, the zero, written the way we do, does fit the angle explanation perfectly.

Questions to Ponder

1. If you were asked why we write the number 6 in that way (or any other number for that matter), how would you respond?
2. Do you remember learning how to write the numbers? What do you remember?
3. How would you teach someone how to write the numbers 0–9?
4. What questions would you ask to help students remember how to write the numbers 0–9?
5. How are the roman numerals and the Arabic numerals related?

Mystery #7

What Does It Mean to Regroup, Borrow, or Carry?

Short Answer

We use a number system based on the number 10. When adding ones, for every ten 1s added, you must add a 1 to the next left column or the "tens" column. When a person adds the tens column for every ten 10s added, one must add a 1 to the next left column, and so forth.

Detailed Answer

Teachers have been showing students how to regroup for years. They call it borrowing or regrouping when doing subtraction and carrying or regrouping when doing addition.

The point of this explanation is to show what is actually going on when we regroup. To understand the concept behind this, one needs to truly understand the concept of place value. When one regroups, borrows, or carries while performing basic operations, one is simply trading. For example, you can trade ten 1s for one 10. You regrouped your 1s into 10s. The best way to introduce this concept is with base ten blocks. But first, let us look at a typical problem and where the mystery seems to appear.

Example: Traditional addition with regrouping

$$\begin{array}{r} 1\ 1 \\ 435 \\ +\ \ 688 \\ \hline 1,123 \end{array}$$

Here is the regrouping. What do the 1s stand for?

The mystery is in the little 1s that seem to appear from nowhere and/or are not given a clear description of what they represent in the procedure. The following explanations will help illustrate just what is going on and what those numbers represent.

Base Ten Blocks

We will take a look at this from a manipulative point of view first. We need to ensure that the reader and the author are on the same page. So we will define base ten blocks.

The following explanation will use base ten blocks. They are used in the primary grades to facilitate the development of place value. There are several types of blocks, and they come in an assortment of colors. The three blocks we will use in this explanation are described below.

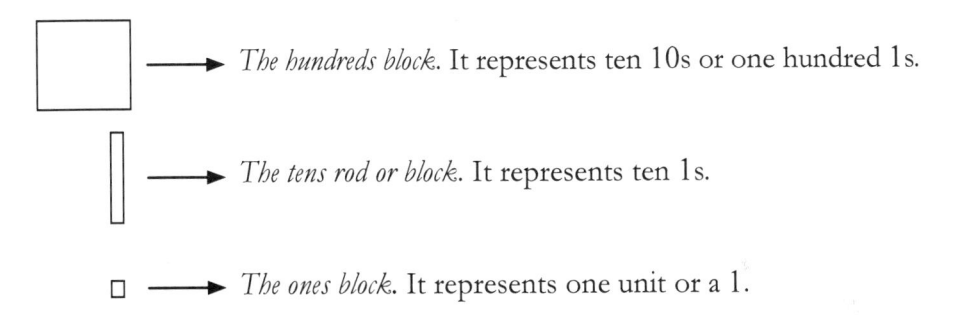

The hundreds block. It represents ten 10s or one hundred 1s.

The tens rod or block. It represents ten 1s.

The ones block. It represents one unit or a 1.

Now that we know what the blocks represent, let's look at the problem mentioned before:

```
  435
+ 688
```

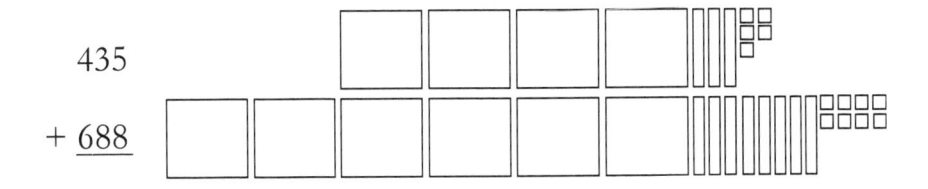

To the right of each number is their base ten block representation. For example, 435 is represented with four *hundreds blocks*, three *tens blocks*, and five *ones blocks*.

To add using the blocks we simply sort them and then count.

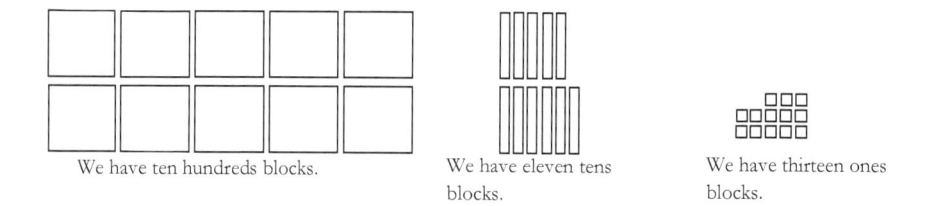

We have ten hundreds blocks.

We have eleven tens blocks.

We have thirteen ones blocks.

Let's follow the standard algorithm for adding these two numbers together.

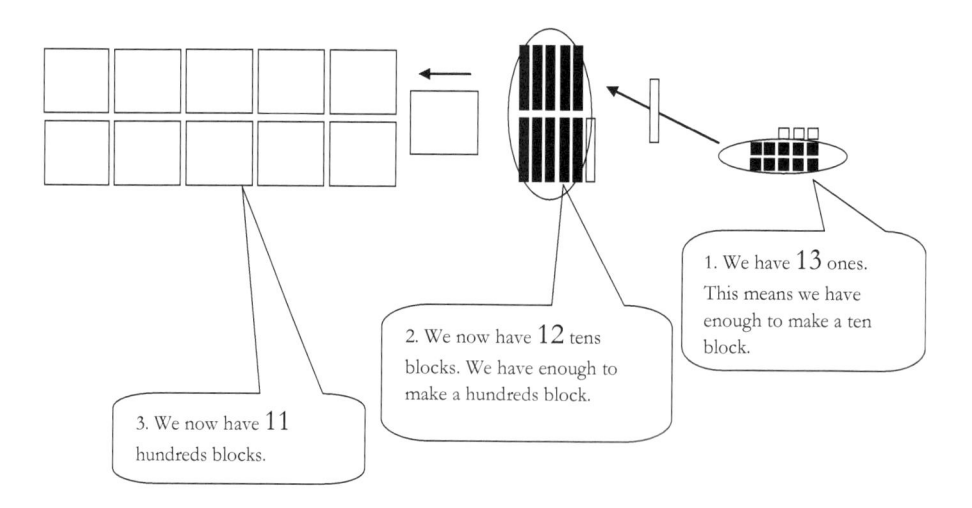

1. We have 13 ones. This means we have enough to make a ten block.

2. We now have 12 tens blocks. We have enough to make a hundreds block.

3. We now have 11 hundreds blocks.

So what do we have now? From the ones column, we have 3 ones left; 10 of them went to the tens column. In the tens column, we have 2 tens left because 10 of them went to the hundreds column. And in the hundreds column, we have 11 hundreds blocks. We could have made a thousands block here, but for now we will simply state our answer as 11 hundreds, 2 tens, and 3 ones, or 1,123.

Shifting Perspective

When teaching this to students, you may find that students would rather add first the hundreds blocks and then the tens blocks and then the ones—in other words, in reverse order than the traditional algorithm; let them! Students with a deep understanding of number sense will be able to add and subtract starting from any place value. It is quite fun. Try it!

Using Expanded Notation

The following is another way to see what is going on. This method uses expanded notation to help visualize the traditional algorithm.

First, we will rewrite the numbers using expanded form:

$$\begin{array}{r} 435 = \\ +\ \underline{688 =} \end{array} \qquad \begin{array}{r} 400 + 30 + 5 \\ +\ \underline{600 + 80 + 8} \end{array}$$

Now let's follow the traditional algorithm to find the sum of the two numbers.

First, we will add the ones column, and notice that since we have $5+8=10+3$ *or* $+3$, we write the 3 in the ones place and regroup the ten.

$$\begin{array}{r} 1 \\ 435 = \\ +\ \underline{688 =} \\ 3 = \end{array} \qquad \begin{array}{r} 10 \\ 400 + 30 + 5 \\ +\ \underline{600 + 80 + 8} \\ 3 \end{array}$$

Now we will add the tens column, and notice that when we add $10+30+80$, we get the sum of 120. That means we need to regroup a hundred.

$$\begin{array}{r} 1\ 1 \\ 435 = \\ +\ \underline{688 =} \\ 23 = \end{array} \qquad \begin{array}{r} 100\ \ 10 \\ 400 + 30 + 5 \\ +\ \underline{600 + 80 + 8} \\ 20 + 3 \end{array}$$

We are now at the final column, the hundreds place. When we add $100 + 400 + 600$, we get $1{,}100$. We need to regroup a thousand.

$$\begin{array}{r} 1\ 1\ 1 \\ 435 = \\ +\ \underline{688 =} \\ 1{,}123 = \end{array} \qquad \begin{array}{r} 1{,}000\ \ 100\ \ 10 \\ 400 + 30 + 5 \\ +\ \underline{600 + 80 + 8} \\ 1{,}000 + 100 + 20 + 3 \end{array}$$

As you can see in this progression, it is much easier to see the mathematics behind the little 1s that we have grown so accustomed to using. The algorithm on the left is much more concise, but the one on the right is less mysterious.

Another Approach to Regrouping and Carrying

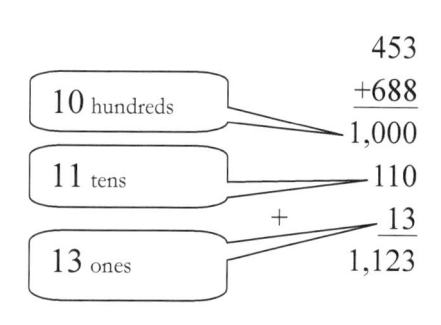

```
          453
         +688
         ----
         1,000
           110
       +    13
         -----
         1,123
```

- 10 hundreds
- 11 tens
- 13 ones

In this example, we are only adding the numbers in each place value. For example, the 4 and 6, which represent 400 and 600, are added together to get 1,000 or 10 hundreds, the 3 and 8 are added together to get 11 tens, and the 5 and 8 are added together to get 13 ones. And then the final sum is achieved by doing the addition to the left without regrouping.

We can also look at regrouping in another way. Let's do the same question but change the format of the regrouping.

| 10
435
+ 688
3 | ←Here we regrouped the 10 from 5 + 8 = 10 + 3 | 100
10
435
+ 688
23 | ←Here we regrouped the 100 from 10 + 30 + 80 = 100 + 20. | 1000
100
10
435
+ 688
123 | ←Here we regrouped the 1,000 from 400 + 100 + 600 = 1,000 + 100. |

Notice that the question is not finished here, but one would only have to bring down the 1 representing how many thousands there are in the sum.

This approach can be used in conjunction with the previous approach to help develop number sense of place value. Place value truly is at the heart of understanding what is going on when regrouping. If place value is properly addressed and understood, then the concept of regrouping is much easier.

Questions to Ponder

1. How would you define regrouping, borrowing, or carrying to someone who does not know what it is?
2. If someone is having a hard time with this algorithm, or set of steps, what might you try?
3. Explain how to subtract the following using base ten blocks:

$$\begin{array}{r} 623 \\ -475 \\ \hline \end{array}$$

4. Explain how to subtract the following using expanded notation:

$$\begin{array}{r} 623 \\ -475 \\ \hline \end{array}$$

5. Show how base ten blocks and expanded notation could be used to illustrate the addition of the following:

$$\begin{array}{r} 12.4 \\ +10.9 \\ \hline \end{array}$$

Mystery #8

Why Can't We Divide by Zero?

Short Answer

First of all, dividing by zero is not equal to zero. It is actually closer to infinity. (I will explain later.) Second, try dividing something into zero "equal" parts. For example, when we divide, let's say, 10 by 2, we are dividing 10 into equal parts or groups of 2 each, and we arrive at 5 groups of 2. Now try dividing the number 10 into equal parts of 0 each. It is crazy to think about it, but it just cannot be done.

Another point of confusion for people is zero divided by something, such as $0 \div 16$. This is zero. Why? Well, if we have zero items and we want to divide these items among 16 people, how many items does each person receive? Right . . . zero.

Detailed Answer

Division by zero is an operation for which you cannot find an answer. In mathematics, when something like this occurs, we often label it as *undefined*, or in other words, $16 \div 0 = $ *undefined*. To help comprehend why this happens, we will first explore the relationship between multiplication and division.

Observe the following example:

We say that $16 \div 8 = 2$ because $8 \times 2 = 16$.

Using this line of reasoning, we could say the following if dividing by zero were permissible:

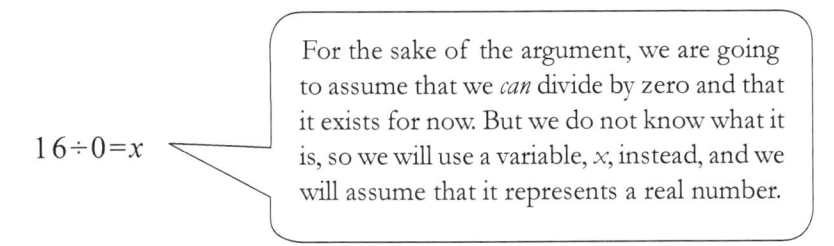

$$16 \div 0 = x$$

> For the sake of the argument, we are going to assume that we *can* divide by zero and that it exists for now. But we do not know what it is, so we will use a variable, x, instead, and we will assume that it represents a real number.

If the above statement were true, then the following statement would also have to be true because of the relationship stated above between multiplication and division.

$$0 \times x = 16$$

This means that there exists some number. We don't know what it is yet or if it even exists, so we've placed an x there that when multiplied by 0 is equal to 16. However, the zero property of multiplication states, "Any number multiplied by zero is equal to zero." (For an explanation on why, please see "Mystery #15: Why is 0 Times Any Number Equal to 0?")

No value would work for x in the expression above because 0 times any number is equal to 0. We can conclude then that division by zero is prohibited.

A Subtraction Approach

We know that one way of looking at multiplication is repeated addition. For example, $2 \times 8 = 16$ is the addition of the number 2 eight times or

$$2+2+2+2+2+2+2+2 = 16.$$

So remembering that multiplication and division are inverse operations and that addition and subtraction are inverse operations, we can look at division as repeated subtraction. We then see that $16 \div 2$ can be rewritten as $16 - 2 - 2 - 2 - 2 - 2 - 2 - 2 - 2$. Subtracting 2 from 16 as many times as we need until we have nothing left results in eight groups of 2. This works with any division problem, except when dividing by zero.

Observe what would happen if we apply the same technique to the following problem:

$$16 \div 0$$

If we do repeated subtraction, $16-0-0-0-0-\ldots$, we would never find out how many groups of 0 are in 16 because, each time, we are subtracting nothing or 0 from 16.

The repeated subtraction approach together with the exploration of the relationship between division and multiplication give a concrete view of the rule that in mathematics, one cannot divide by zero. Doing so gives the students an understanding of the reason why one cannot divide by zero instead of giving a baseless answer, which leaves the student in the dark and creates a mystery around this mathematical concept.

Interesting Observation

Many students assume that when dividing by zero, the answer is zero. So a subsequent question to the mystery we answered above could be,

Why isn't a number divided by zero equal to zero?

We will take 16 and divide it by different numbers to see if we can see a pattern and determine what the answer would be if we could divide by zero.

$$16 \div 16 = 1$$
$$16 \div 8 = 2$$
$$16 \div 4 = 4$$
$$16 \div 2 = 8$$
$$16 \div 1 = 16$$
$$16 \div 0.5 = 32$$
$$16 \div 0.25 = 64$$
$$\cdot$$
$$\cdot$$
$$\cdot$$
$$16 \div 0.000003814697266 = 4{,}194{,}304$$

There are a few things that we should notice here.

1. As the divisors decrease, the quotients (or answers) increase.
2. The divisors are getting closer and closer to zero.
3. If we divided 16 by a divisor that was really close to zero, the quotient would be really big.

Go ahead and try it: divide any number by a number so small that it is really close to zero (see glossary: *infinitesimal*) and predict how big the number would be.

One can easily see that if this pattern were to continue, the quotient would continue to increase until it would reach an undefined amount. In the world of mathematics, we usually refer to this amount as infinity.

Questions to Ponder

1. Why can't we divide by zero?
2. How would you explain the answer to $18 \div 0$?
3. Compare the following:

$$\frac{0}{4} \quad \& \quad \frac{4}{0}$$

   Which one is bigger? Which one is smaller? In which context would you have one, and which one would you have the other?

4. How would you explain $\frac{0}{0}$?

Mystery #9

Why Do We Do It That Way? Long Division

Short Answer

Most students love or at least tolerate mathematics during elementary school, but then somewhere, they come to one of the great stumbling blocks—long division. Many people have tried to push long division out of education, claiming that with the invention of the calculator and computer, this algorithm is no longer needed. Without going into too much history, we can simply say that it is still with us, and with the advent of the new Common Core State Standards, it will be with us in the future as well. So in short, long division is the process for dividing numbers, usually when the divisor is a multidigit number, like 435 divided by 12 or $12\overline{)435}$.

Detailed Answer

Long division is difficult for students. There are a lot of steps. Also, long division is the only arithmetic operation that traditionally starts in the leftmost place value. All others start in the ones place. Right away we are in a different and potentially confusing territory for most students.

Another reason long division might be difficult is because the students have yet to master their multiplication tables. This is a tremendous obstacle for students in many concepts, but it doesn't have to prevent them from mastering long division without knowing their multiplication tables.

We will first go through the steps for the algorithm and look more deeply at what is going on, and then we will make some modifications as to how it is taught.

Let's start with a basic problem.

$$12\overline{)358}$$

I will now attempt to duplicate what a teacher says and does to teach this concept to students. If I seem condescending, I apologize, but it can't be helped.

$$12\overline{)358}$$

The first thing that teachers usually say is, "How many times will 12 go into 3?"

Since 12 doesn't go into 3, we move over one number and now ask the question again, "How many times does 12 go into 35? Sometimes they add to it, "How many times does 12 go into 35 without going over?" And then a side note, "We have to get as close as we can!" OK, 12 goes into 35 two times because three times is too much.

Then we *multiply*!

Yep, you heard me right. This is a division problem, but we are multiplying.

Now 12 times 2 is 24. We then write the 24 under the 35 and then *subtract*!

OK, we need a visual of what is going on. Here is what we have so far:

$$\begin{array}{r} 2 \\ 12\overline{)358} \\ \underline{24} \\ 11 \end{array}$$

Yes, it looks like quite a mess, but trust me, it gets worse!

Now you see that 8 just sitting there? We must bring it down! Here is a visual of what I mean:

$$\begin{array}{r} 2 \\ 12\overline{)358} \\ \underline{24}\downarrow \\ 118 \end{array}$$

Teachers will now start the process all over again. "Does the 12 go into the 1, etc.?"

Let's go ahead and finish the problem so we can end the torture. Here is the problem in its finished glory. We will leave it with the remainder written at the bottom.

$$\begin{array}{r} 29 \\ 12\overline{)358} \\ \underline{24}\downarrow \\ 118 \\ \underline{108} \\ 10 \end{array}$$

So 12 divides 358 a total of 29 times with 10 left over. Sometimes we write it as 29 R 10.

The unfortunate thing is that I do not think I exaggerated much on the explanation. Teachers will go over the explanation much like what we saw and then have the students practice those steps. Students will be able to mimic the steps, but they often get the steps mixed up. They will forget to bring a number down or add instead of subtract, etc.

Unraveling the Mystery

To unravel this mystery we need to look at it from a different point of view. We will remember that we have chosen to see division as repeated subtraction throughout many of the sections in this book. So we will look at the previous problem through the lens of repeated subtraction.

$$12\overline{)358}$$

How many times can we subtract 12 from 358? We could do it one 12 at a time as shown below, but it might take an awful long time.

$$\begin{array}{r} 12\overline{)358} \\ -12 \\ \hline 346 \\ -12 \\ \hline 334 \\ \vdots \end{array}$$

Even though this may take a while, it might be beneficial to go through the tediousness of the act to help students see a shortcut. Let's jump to the shortcut.

After subtracting 12 several times, many should realize that it is possible to subtract bunches of 12 at the same time. For example, we can subtract a group of ten 12s at a time. See below.

$$\begin{array}{r} 12\overline{)358} \\ -120 \\ \hline 238 \\ -120 \\ \hline 118 \end{array}$$

In this case, we have subtracted 120 twice so far. As shown above, this is more efficient than subtracting one 12 at a time.

We should finish the problem now. We can't subtract any more 120s, but we can subtract a smaller number, such as 60 (which is half of 120, or 12×5). See below.

$$
\begin{array}{r}
12\overline{)358} \\
-120 \\
\hline
238 \\
-120 \\
\hline
118 \\
-60 \\
\hline
58
\end{array}
$$

We cannot subtract 60 anymore, but we can subtract 24 ($12 \times 2 = 24$).

$$
\begin{array}{r}
12\overline{)358} \\
-120 \\
\hline
238 \\
-120 \\
\hline
118 \\
-60 \\
\hline
58 \\
-24 \\
\hline
34 \\
-24 \\
\hline
10
\end{array}
$$

We have arrived at a remainder of 10. This is what we arrived at when we did the traditional algorithm, but we didn't keep track of what we subtracted. Let's review.

First we subtracted two 120s, or two bunches of 12×10s. That was 20 twelves. Then we subtracted one 60, or one bunch of 12×5s. That was 5 twelves. Finally, we subtracted two 24s or two bunches of 12×2s. That was 4 twelves. We have subtracted 29 twelves from 358 with a remainder of 10.

The traditional algorithm hides all the mathematics we just did. We have not explicitly explained the mystery behind the long division algorithm, but we have shown the mathematics behind it.

Why Don't We Start in the Ones Place?

As was mentioned above, this operation does not have students starting in the ones place like the other operations. A quick explanation of this is that it is more efficient to divide up the bigger numbers first than the smaller ones.

It might be easier to explain this with base ten blocks. Let's use the same problem we explored above.

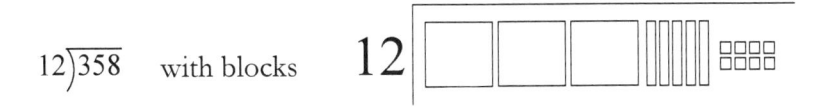

$12\overline{)358}$ with blocks

To divide 358, we will start with the hundreds. But we don't have 12 of them to divide up, so we must convert them to 10s. On the left we will show long division with its traditional steps, and on the right we will show what is happening with the blocks.

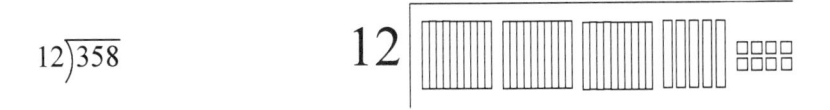

$12\overline{)358}$

It is now easier to divide up the hundreds. Notice that we now have 35 hundreds to divide up. There is no real easy way to do this, so I will have to try and do this as clearly as possible. The illustration below will show better the next step in the algorithm.

$$\begin{array}{r} 2 \\ 12\overline{)358} \\ \underline{24} \\ 11 \end{array}$$

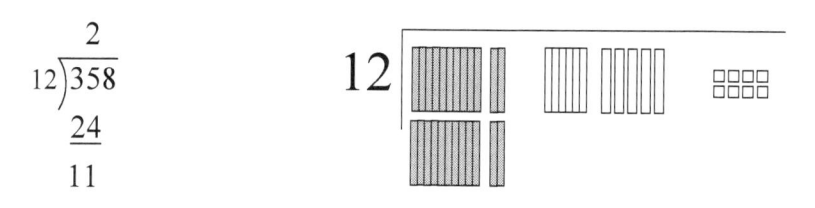

On the left are some shaded tens blocks. These are the blocks that have been divided up into groups of 12. Notice that there are two of them and that there are eleven 10s left after grouping them together. This is what is represented with the long division algorithm shown on the left. Let's continue, noting that since we cannot divide the 10s up any longer that they must be broken down into 1s.

$$\begin{array}{r} 29 \\ 12\overline{)358} \\ \underline{24}\downarrow \\ 118 \\ \underline{108} \\ 10 \end{array}$$

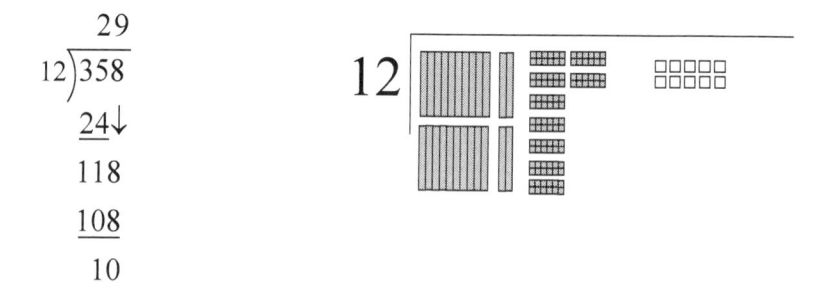

Notice that we now have another set of grouped blocks, this time of 1s. To be exact, we have 9 groups with 10 remaining. This, again, coincides with what is represented in the long division algorithm on the left.

If we look now at the groups of 12 that we have made, we have 2 groups of tens and 9 groups of ones, or 29 with 10 left over.

Questions to Ponder

1. How would you define long division to someone who does not know?
2. Explain how to do the following problem with base ten blocks:

$$12\overline{)579}$$

3. Do you remember learning long division? How was it taught to you?
4. With all the technology available, do you think it necessary to have students learn the long division algorithm? Why or why not?
5. Research some of the other ways to do long division. How are they related to algorithm shown above?

MYSTERY #10

Why Do We Do It That Way?
"Invert and Multiply"
when Dividing by a Fraction

Short Answer

The most popular answer is that it is easier to invert and multiply than to divide with fractions, especially if one looks at division as repeated subtraction. In general, operations with fractions are considered difficult procedures to execute for students, but they can be made easier if we sell them the need for the algorithm in the first place.

Detailed Answer

Even though we are going to find the reason why we do the whole "invert and multiply" procedure, we will also try and attempt to explain what is going on when dividing by a fraction.

This is one of those long-standing mysteries. Elementary and secondary teachers alike remember being taught the algorithm but don't ever remember being taught the reason why. Most teachers simply remember the old saying, "Ours is not to reason why, just invert and multiply." As a result, this is one of the major mysteries of mathematics for students. Let us first take a look at the standard algorithm for dividing by a fraction.

$$4 \div \frac{1}{2} = \frac{4}{1} \times \frac{2}{1} = \frac{8}{1} \text{ or } 8$$

Above is the traditional way to divide by a fraction, in this case $4 \div \frac{1}{2}$. The steps are usually explained this way: first, change the division sign to a multiplication sign, then flip the second fraction, and finally, multiply.

The mystery is that students are never taught or never allowed to explore the mathematics behind the middle step. Let's see if we can look at what is going on in the division by a fraction to develop some understanding of the algorithm.

Let's say we are doing the following operation: $4 \div \frac{1}{2}$

$4 \div \frac{1}{2}$ rewritten as $\dfrac{4}{\frac{1}{2}}$.

> This still means $4 \div \frac{1}{2}$. We are just using a *fraction bar* instead of the *division sign*.

Now we want to clear out the fraction in the denominator so that we only have a fraction in the numerator and not in the denominator. To do this, we can multiply the denominator by its reciprocal as follows:

$$\dfrac{4 \times \frac{2}{1}}{\frac{1}{2} \times \frac{2}{1}}$$

> Notice, we also multiplied the numerator by $\frac{2}{1}$ to conserve the value of the expression. It is the old "what we do to the denominator, we must do to the numerator" concept, which preserves the value of the expression.

Now we can simplify our problem a little and get the following:

$$\dfrac{4 \times \frac{2}{1}}{1} \text{ or } 4 \times \frac{2}{1}$$

> If you compare what we have now with what we started with, we have inverted the second fraction, and our problem has gone from a division problem to a multiplication problem. We have essentially inverted and multiplied.

59

Another Approach

The above approach used a lot of manipulation and algebra. Let's see if we can use the number line to illustrate this. Let's take the same problem:

$$4 \div \frac{1}{2}$$

But place 4 on the number line and use repeated subtraction, in this case subtract $\frac{1}{2}$ from 4 as many times as we can until we reach zero.

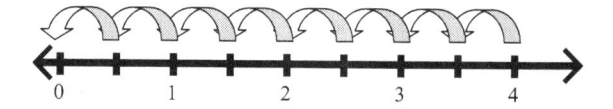

Each jump above represents going back exactly $\frac{1}{2}$. As we jump back $\frac{1}{2}$, we are subtracting $\frac{1}{2}$ each time. So by the time we have arrived at the number 0, we have subtracted $\frac{1}{2}$ eight times. In this manner, we can more easily see what dividing by a fraction does and why the flipping of the fraction is necessary.

Questions to Ponder

1. How would you explain why we invert and multiply when dividing by a fraction to someone who would ask?
2. Instead of remembering "Ours is not to question why, just invert and multiply," is there another way of remembering the steps that does not suggest we shouldn't question why?
3. How would you explain $13 \div \frac{1}{2}$ using a number line?
4. Come up with a story problem for the following number sentence:

$$15 \div \frac{3}{4} = 20.$$

5. Write out a real-world context for the following number sentence:

$$\frac{3}{4} \div \frac{1}{2} = 1\frac{1}{2}.$$

Mystery #11

Why Do We Do It That Way? Converting a Mixed Number to an Improper Fraction

Simple Answer

A mixed number is made up of a whole number part and a fraction part. Converting a mixed number into an improper fraction simply converts the whole number into its fractional equivalent and then adds to the remaining fractional part.

Detailed Answer

That simple answer probably didn't cut it for a good explanation. So I am glad you have continued to read. Let's see if we can help you understand it a little better.

Fractions have traditionally been one of the most difficult concepts for students to work with in mathematics. There have been many debates on the reasons why students fail to grasp fractional understanding, but most agree that students lack a key foundational point in the concept of fractions.

One of the purposes of this explanation is to assist a student's understanding of mixed numbers and improper fractions so that the student can connect the routine with the concept.

Let's explore the mixed number $2\frac{1}{3}$. The traditional technique for converting a mixed number to an improper fraction is to multiply the whole number by the denominator of the fractional portion and then adding that product to the numerator.

See below:

$$2\frac{1}{3} = \frac{2 \times 3 + 1}{3} = \frac{7}{3}$$

Now there are several ways of looking at this technique. Let's first look at it strictly from a computational point of view. The fraction $2\frac{1}{3}$ is usually read as "two and one-third." We could use this and rewrite what we said with math symbols (remember the word *and* is commonly translated with a +) and so $2+\frac{1}{3}$, which has now become an addition problem. See below as we do the addition of these two numbers using a standard algorithm.

$$2+\frac{1}{3} \rightarrow \frac{2\times 3}{3}+\frac{1}{3} \rightarrow \frac{2\times 3+1}{3} \rightarrow \frac{7}{3}$$

or

$$\frac{3}{3}+\frac{3}{3}+\frac{1}{3}=\frac{7}{3}$$

Notice two things:
1. We arrived at the same answer, $\frac{7}{3}$.
2. We are essentially doing the same thing as the traditional method.

In the latter method, when we changed the whole number to a fraction with a common denominator, we had to multiply the numerator, in this case 2 by 3. Then we added the two numerators together to arrive at the desired sum.

This approach does not always convince students and teachers, so we will attempt another approach. On the next page we will attempt to illustrate the procedure with a diagram.

Graphical Approach

Another way to look at this is to use manipulatives. Let's look at $2\frac{1}{3}$ again, except this time with blocks.

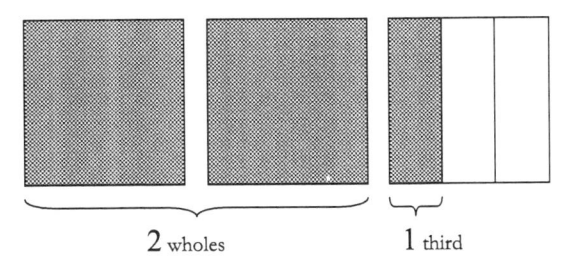

<u>2 wholes</u>   <u>1 third</u>

Now break the 2 wholes up into thirds.

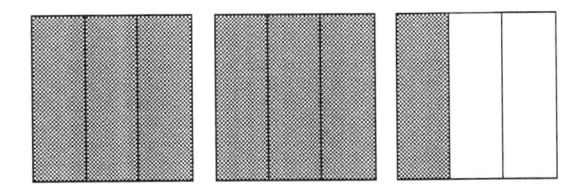

If we count the number of thirds, we find that we have "seven thirds" or $\frac{7}{3}$.

We can also see that our two operations are still present in this visual.

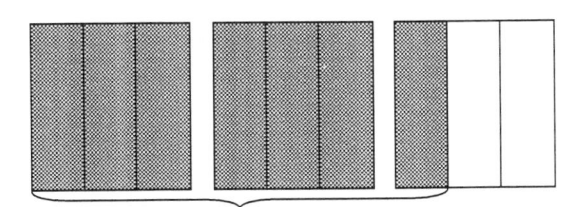

2 groups of 3 "thirds" + 1 "third" = 7 "thirds" or

$$\frac{3}{3} + \frac{3}{3} + \frac{1}{3} = \frac{7}{3} \text{ or } \frac{2 \times 3 + 1}{3} = \frac{7}{3}$$

Students and teachers appreciate this particular method. Once they see the mathematics behind the procedure, many, if not all, will always remember how.

Questions to Ponder

1. How would you explain the procedure for converting a mixed number to an improper fraction to someone who would ask?
2. Explain the conversion of $3\frac{1}{5}$ to an improper fraction using algebra.
3. Explain the conversion of $3\frac{1}{5}$ to an improper fraction using a drawing.
4. Why do we convert mixed numbers to improper fractions?
5. Which are easier to work with, improper fractions or mixed numbers? Why?

Mystery #12

Why Do We Do It That Way? Converting an Improper Fraction to a Mixed Number

Short Answer

An improper fraction is a fraction where the numerator is greater than the denominator. Converting an improper fraction into a mixed number is simply dividing the numerator by the denominator. The result is usually a whole number part and the remainder is converted into a fraction.

Detailed Answer

Again, much like the related topic that preceded this one, it is probably good that you have continued to read. Let's see if we can make sense of it here.

The same set of manipulatives can be used to illustrate converting an improper fraction to a mixed number. Before that, however, I would like to address one of the reasons why students forget how to do this operation or algorithm.

$\frac{7}{3}$ — This line means division. We could rewrite this division problem as 7÷3 or "7 divided by 3" and then naturally write $3\overline{)7}$ and perform the basic algorithm of which we are accustomed to arrive at the desired result (a mixed number).

The traditional algorithm performed looks like this:

$$3\overline{)7} \quad \begin{array}{r} 2 \\ \underline{-6} \\ 1 \end{array}$$ and then made into a mixed number $2\frac{1}{3}$.

Now let's look at it with manipulatives.

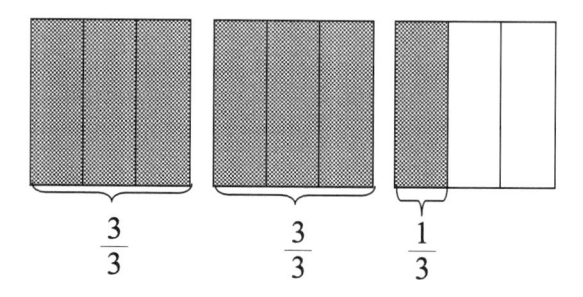

Let's remember that $\frac{3}{3} = 1$. Therefore, we have

$$\frac{3}{3}+\frac{3}{3}+\frac{1}{3} \to 1+1+\frac{1}{3} \to 2+\frac{1}{3} \to 2\frac{1}{3}.$$

Together with the preceding mystery, we have a couple of the most "mysterious" mysteries of all. The reason is probably because it is fairly easy to practice with the students and teach. However, I have found that many students forget the steps for the conversion, and giving them a reason is very beneficial.

More importantly, revealing the mathematics behind converting an improper fraction to a mixed number and vice versa helps students comprehend their values more completely. Often, students misunderstand their values, especially when asked to compare and order numbers, such as fractions and decimals.

Questions to Ponder

1. How would you explain the procedure of converting an improper fraction to a mixed number to someone who would ask?

2. Why do we convert improper fractions to a mixed number?
3. Explain the operation $\frac{12}{7}$ using algebra.
4. Explain the operation $\frac{12}{7}$ using a diagram.
5. Are there situations in which using improper fractions is preferred over mixed numbers?

Mystery #13

Why Do We Do It That Way? Counting Decimal Places when Multiplying with Decimals

Short Answer

When someone is multiplying two numbers and one or both contain decimals, they are essentially multiplying mixed numbers. Decimals are just another form of writing mixed numbers, and when we multiply mixed numbers, we must multiply the fractions as well as the whole numbers. When we count the decimal places after multiplying, we are taking into account that we are not multiplying whole numbers but mixed numbers with a fractional part.

Whoa! That was very busy. Let's try looking at this step-by-step.

Detailed Answer

In other words, after multiplying two numbers (in the form of decimals) together and before we have our final answer (product), why is it necessary to count the number of decimal places to figure out where the decimal point goes in the product? For example,

$$\begin{array}{r} 0.5 \\ \times 0.3 \\ \hline \end{array}$$ If we go through the multiplication, we get $$\begin{array}{r} 0.5 \\ \times 0.3 \\ \hline 0.15 \end{array}$$.

> 0.5 and 0.3 each have a decimal place giving us two decimal places.

To know that the product has two decimal places, we count the number of decimal places in the initial problem, in this case two. But why do we count the decimal places? Why is the answer "fifteen hundredths," or 0.15?

To truly understand what is going on, let's look at the actual numbers we are multiplying.

0.5, which is read as "five tenths," can be rewritten as $\frac{5}{10}$; and

0.3, which is read as "three tenths," can be rewritten as $\frac{3}{10}$.

If we multiply these two numbers as fractions instead of decimals, we get $\frac{5}{10} \times \frac{3}{10} = \frac{15}{100}$ for which the answer is read as "fifteen hundredths," "one and a half tenths," or 0.15.

Manipulative Approach

We can also illustrate this with a simple tenths-and-hundredths grid. First, let's preface this visual with the fact that multiplication can also be seen as finding the area of a polygon, in this case a rectangle.

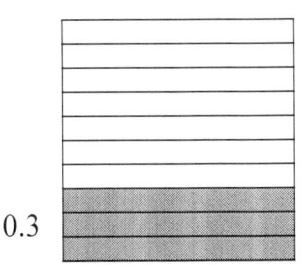

The grid to the left is a tenths grid showing 0.3, or three tenths.

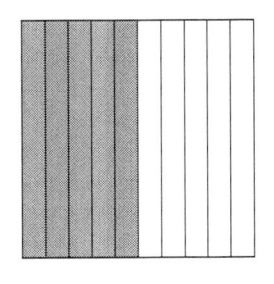

0.3

The grid to the left is a tenths grid showing 0.5, or five tenths. This is sideways for a reason you will see shortly.

0.5

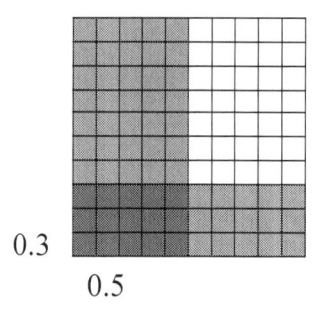

The darker rectangle represents the overlapping area of the 0.5 and the 0.3, or the product of 0.3 × 0.5. There are 15 boxes in all that are overlapping in what has now become a hundredths grid, 0.15, or "fifteen hundredths."

0.3

0.5

The darker-shaded rectangle has an area of 0.15. It has a length of 0.5 and a width of 0.3. The units, hundredths, are easily seen in the visual above when compared to the percentage area of a hundredths grid.

Questions to Ponder

1. How would you explain the reason behind counting the decimals before arriving at the final answer in multiplying decimals?
2. Explain an alternative way to arriving at the answer for 2.3 × 0.31 other than the standard algorithm.
3. Compare

$$\begin{array}{r}0.3\\ \times\,0.5\\ \hline\end{array} \text{ and } \frac{3}{10} \times \frac{5}{10}.$$

4. Write out a real-world context for the following number sentence: 2.5 × 1.5 = 3.75.
5. To do the following problem, does one have to convert both to either fraction or decimal form, or could you do it without converting? Describe what is going on.

$$2.5 \times \frac{3}{4}$$

Mystery #14

Why Do We Do It That Way? Moving the Decimal Point in the Divisor when Dividing by a Decimal

Short Answer

Simply put, it is much easier to move the decimal point over than to do the arithmetic with the decimal. Remember that division can be seen as repeated subtraction. So it is possible to do the division without moving the decimal point. You may just have to subtract a very small number, the decimal, a large number of times. Try it.

Detailed Answer

This algorithm beats a lot of students up. Not only is this long division, but it is long division with decimals! Let's see if we can make this easier.

Let's look at a long division problem with a decimal as a divisor and see if we can find a reason for moving that elusive decimal point. We will use the problem

$$40 \div 0.5.$$

Here is a review of the traditional technique for finding the quotient when dividing by a decimal.

$0.5 \overline{)40}$ To start this problem, we must move the decimal to the right until we have a whole number in the divisor. The number of digits we move the decimal point in the divisor is the number of digits we move the decimal point in the dividend.

$$5\overline{)400.}$$
Notice, now we have 400 ÷ 5 because we moved the decimal one place in the divisor and, therefore, one place in the dividend.

If we go ahead and finish the process, we get the following result.

$$\begin{array}{r} 80. \\ 5\overline{)400.} \\ \underline{40} \\ 00 \\ \underline{00} \\ 0 \end{array}$$

The resulting quotient is 80.

Small note: Remember that division is repeated subtraction. So if we are to completely understand this result, we may see it as subtracting 0.5 from 40 eighty times, which would get us to zero.

To illustrate the reason for moving the decimal point, we will change the deciml to a fraction:

$$400 \div 0.5 \quad \text{will become} \quad 40 \div \frac{5}{10}.$$

We will now apply the traditional algorithm—invert and multiply (please see the explanation in "Mystery #10")—to find the quotient.

$$40 \div \frac{5}{10} \quad \text{becomes} \quad \frac{40}{1} \times \frac{10}{5}, \quad \text{which becomes} \quad \frac{400}{5}, \text{which}$$

we can rewrite as $400 \div 5$ or $5\overline{)400}$, giving us a quotient of 80.

This is the same quotient we got when we moved the decimal place in the divisor using the traditional rule shown above.

When we divide by decimals and move the decimal point over in both the divisor and dividend, we are accounting for the fact that we are essentially dividing by a fraction.

The above explanation is very abstract. There are no visuals to speak of. So my question for you is this: can we come up with an explanation that introduces visuals? For example, since division and multiplication are inverse operations, could we use the same visual we used in the unraveling of that mystery to further unravel this mystery? Let's give it a try.

In the case of the problem at hand, $40 \div 0.5$, we must understand what we are dividing. For example, we are dividing by five-tenths, so our question is, how many tenths are there in 40? And subsequently, how many groups of 5 tenths are there in that many tenths?

How many tenths are in 40? The answer is 400. How did we arrive at that answer? Below is a diagram of ten-tenths or $\frac{10}{10}$, which makes one whole.

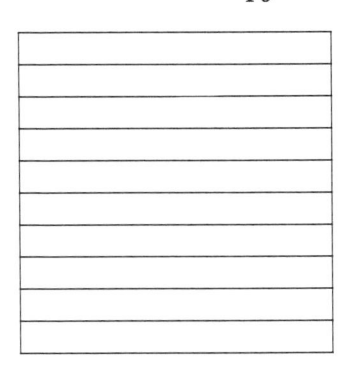

Since we have 40 "wholes," we have forty of these grids. We can multiply the number of grids we would have by the number of tenths in each grid to give us the number of tenths in 40. For example, $40 \times 10 = 400$.

How does this help? Well, now we want to divide each whole by five-tenths. In each tenths grid, we can subtract (remember, division is repeated subtraction) two groups of five-tenths. If we do this for all 40 tenths grids, we will do this a total of 80 times.

Questions to Ponder

1. How would you explain the procedure for dividing by a decimal?
2. Explain the process for $1.3 \overline{)220}$ and make sure you give a reason for moving the decimal.
3. Why do we move the decimal point?
4. Perform the following operation, $10 \div 0.2$, without moving the decimal point.
5. Write a real-world context for the following number sentence:

$$16 \div 0.25 = 64$$

Mystery #15

Why Is 0 Times Any Number Equal to 0?

Short Answer

Remember that multiplication is repeated addition, so 0×0 can be seen as adding zero to itself zero times. This, hopefully, can be plainly seen as equaling zero.

Detailed Answer

We will attempt to show this using a pattern approach. We will use the product of 4×4 to show the pattern. We will decrease both factors until we reach zero to show the pattern. Here we go.

$$4 \times 4 = 16$$
$$3 \times 3 = 9$$
$$2 \times 2 = 4$$
$$1 \times 1 = 1$$
$$0 \times 0 = ?$$

What replaces the question mark? If we look at the products (the answers to each multiplication problem) above, we might see a pattern. The products are as follows:

$$16, 9, 4, 1, ?$$

If we look at the differences between the numbers, we will detect that these differences have a pattern.

Between 16 and 9, the difference is 7, or 16-9=7.
Between 9 and 4, the difference is 5, or 9-4=5.
Between 4 and 1, the difference is 3, or 4-1=3.

If this pattern were to continue, then we would next have a difference of 1, or 1-1=0.

In other words, $0 \times 0 = 0$

Remember, we have not proven anything, only showed through patterns that 0×0 is equal to 0.

Questions to Ponder

1. How would you explain why zero times anything is zero?
2. Draw a diagram that would illustrate what 0×0 would look like.
3. How would you explain $0 \times 0 \times 0$?
4. Write out a real-world context for the following number sentence:

$$15 \times 0 = 0.$$

5. Write out a real-world context for the following number sentence:

$$0 \times 15 = 0.$$

Mystery #16

Why Is the Product of a Positive Integer and a Negative Integer Equal to a Negative Integer?

Short Answer

If we try and make sense of it intuitively, then it isn't too hard to see why. For example, $4 \times (-1)$ can be seen as (-1) added to itself four times, or $(-1)+(-1)+(-1)+(-1)$, which results in -4. This does not prove the point, but if you try this with other numbers, you will see that it also works.

Detailed Answer

A Pattern Approach

A much more in-depth way of explaining this to kids and one that students tend to trust more is through the use of patterns. Remember that one of the definitions for mathematics is the search for patterns.

We will use the product of 4×4 to show the pattern. We will decrease the second factor by one and get a new product each step, working our way down to a negative factor.

$$4 \times 4 = 16$$
$$4 \times 3 = 12$$
$$4 \times 2 = 8$$

$$4 \times 1 = -4$$
$$4 \times 0 = 0$$
$$4 \times (-1) = ?$$

What value should go where the question mark is? If we look at the products (the answers to each multiplication problem) above, we might see a pattern. The products are as follows:

$$16, 12, 8, 4, 0, ?$$

There is a pattern in these numbers: each number is four less than the previous number. For example, 12 is $16 - 4$, and 4 is $8 - 4$, and so forth. We could follow this pattern to uncover the number that goes in the place of the question mark.

The question mark is going to be four less than zero, or $0 - 4 = -4$. And if we take that product and place it back into our set of multiplication facts that we listed off in the beginning, we can see that $4 \times (-1) = -4$.

Therefore, we can see that the product of a positive factor (4) and a negative factor (–1) will result in a negative product (–4).

Questions to Ponder

1. How would you explain why a positive integer times a negative integer is a negative integer?
2. Explain the product of $5 \times (-3)$ without just stating the rule "A positive times a negative is a negative."
3. How do you remember the rule for multiplying a negative integer and a positive integer?
4. Explain this mystery with manipulatives or a drawing.
5. Relate the fact that a positive times a negative integer is a negative integer to the coordinate plane.

Mystery #17

Why Is the Product of a Negative Integer and a Negative Integer Equal to a Positive Integer?

Short Answer

Let's take the product of −3 × (−2). We can try the reasoning we used before. Multiplication is repeated addition. This means that −3 × (−2) is −3 added to itself (−2) times. This is not as intuitive as the previous question, at least not with repeated addition.

There are a lot of cute ways of showing that the product of a negative integer and another negative integer is equal to zero. Some explanations involve money or sliding an object forward and backward along a number line. These explanations have some merit. We will attempt to look at this rule in a different way.

Detailed Answer

We will use the same thinking pattern used in the previous section.

We will begin where we left off the last set of patterns and see what happens when we multiply a negative number with a negative number.

$$3 \times (-1) = -3$$
$$2 \times (-1) = -2$$
$$1 \times (-1) = -1$$

$$0 \times (-1) = 0$$
$$-1 \times (-1) = ?$$

What value should go where the question mark is? If we look at the products (the answers to each multiplication problem) above, we might see a pattern. The products are as follows:

$$-3, -2, -1, 0, ?$$

There is a pattern in these numbers: each number is one more than the previous number. For example, -2 is $-3 + 1$, and -1 is $-2 + 1$, and so forth. We could follow this pattern to uncover the number that goes in the place of the question mark.

The question mark is going to be one more than zero, or $0 + 1 = 1$. And if we take that product and place it back into our set of multiplication facts that we listed off in the beginning, we can see that $-1 \times (-1) = 1$.

Therefore, we can see that the product of a negative factor (-1) and a negative factor (-1) will result in a positive product (1).

This line of thinking can be used with any two sets of numbers and show the same result. One of the benefits to showing the rule this way is that it takes away the mystery as well as gives exposure to the type of thinking students will need to figure things out on their own later.

Questions to Ponder

1. How would you explain why a negative integer times a negative integer is a positive integer?
2. How do you remember the rule for multiplying a negative integer and a negative integer?
3. Explain the product of $(-5) \times (-3)$ without using the phrase "A negative times a negative is a positive."
4. Explain this mystery with manipulatives or a drawing.
5. Relate the fact that a negative times a negative integer is a positive integer to the coordinate plane.

Mystery #18

Why Is Any Number Raised to the Zero Power Equal to One, or in Other Words, Why Is $n^0 = 1$?

Short Answer

The quickest and easiest way to show this rule to students is to show by a pattern. Let's look at the following list of products:

$$3^3 = 27$$
$$3^2 = 9$$
$$3^1 = 3$$
$$3^0 = ?$$

We notice two patterns in the list above: (1) the exponents decrease by one, and (2) the products decrease by a factor of 3, or each succeeding product is the previous product divided by 3 (i.e., we get 9 by dividing 27 by 3, and so forth). So to find the value that replaces the question mark, we can take the previous product, in this case 3, and divide it by 3. The result is 1.

Detailed Answer

"Anything raised to the zero power is equal to one." Teachers tell this to students all the time but fail to give any explanation for this. They usually just give a bunch of examples to establish the rule.

For example, teachers may write on the board the following:

$$3^0 = 1$$
$$12^0 = 1$$
$$100^0 = ?$$

Simply giving students examples does not give students comprehension and understanding to the mathematics behind the rule.

This mystery can be explained easily in two ways. We will first explore this with patterns, and then we will show this graphically.

A Pattern Approach

We will choose a small value for our base, 2, so that we can illustrate this concept without going into large numbers.

Look at the following set of powers of 2. Notice that if you start at the top of the sequence and then proceed down, the values of the products are found by dividing the preceding value by 2.

$$2^4 = 16$$
$$2^3 = 8$$
$$2^2 = 4$$
$$2^1 = 2$$
$$2^0 = ?$$

We will attempt to establish some sort of pattern in the products to determine what should go where the question mark is, assuming we don't know yet.

The products are as follows:

$$16, 8, 4, 2, ?$$

Each number is half of the preceding number. For example, the 8 is 16 ÷ 2, and the 4 is found by dividing the 8 by 2.

If we were to follow this pattern to the location of the question mark or the value of 2^0, we would take the preceding product, or 2, and then divide it in half, or in other words, 2 ÷ 2 = 1.

We can now see that $2^0 = 1$.

A Graphical Approach

Now let's see what this would look like graphically. I have attempted to make the graph as clear as possible. Below is a graph that illustrates the pattern that was shown above.

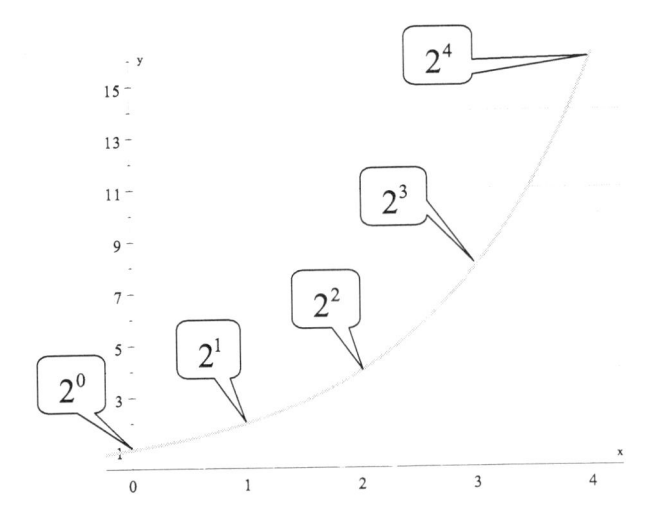

We can see by this graph that the above pattern creates an obvious curve for the results.

Notice that as the curve moves from right to left that it does not approach 0, but it crosses through 1.

Thus, we again see that $2^0 = 1$.

The graphical approach will rely on some assumption of some student understanding of graphing. The major benefit of the graphical approach is that students trust pictures. If a teacher can show that something works with a picture, then the understanding and retention of the rule by the student is increased.

Another way to improve this method would be to have the students duplicate the graph with a base of another value, such as three.

Questions to Ponder

1. How would you explain why any number raised to the zero power is equal to 1?
2. Explain the equation $5^0 = 1$ without just stating examples.
3. How do you remember that any number raised to the zero exponent is 1?
4. Write out a real-world context for the following equation:

$$2^0 = 1$$

5. Is there a situation in mathematics where a number raised to an exponent equals to zero? Explain why or why not.

Mystery #19

Why Is Any Number Raised to a Negative One Equal to Its Reciprocal, or in Other Words, Why Is $n^{-1}=\frac{1}{n}$?

Short Answer

Before we continue, we should note that this explanation and the last one go together, and to truly benefit from this explanation, you should read both. So if you haven't, we'll wait as you do so.

OK, now that you are back with some background knowledge, we may continue.

Let's remember that the exponent is used as reference and not as a multiplier. Remember that $3^2 \neq 3 \times 2$, but it does mean 3×3.

What implications does that have on the negative exponent? What does 3^{-1} mean?

First, we can say that $3^{-1} \neq 3 \times (-1)$.

With the definition of an exponent that we talked about, we can state that 3^{-1} means to multiply 3 by itself negative one time or to multiply 3 by itself the opposite of one time. What is the opposite of multiplying something by itself? The answer is division. So instead of multiplying 3 by itself once, we are asked to divide it by itself once. When we multiplied 3 by itself once, we simply arrived at the answer of 3. It was kind of like multiplying 3 by 1. In this case, we are now dividing 1 by 3, or in other words,

$$1 \div 3 = \frac{1}{3}.$$

A negative exponent can also be looked upon as meaning "the reciprocal of." For example, 3^{-1} can be read as "the reciprocal of 3," or $\frac{1}{3}$.

Detailed Answer

Much like the previous mystery, we will take two approaches to this mystery. The first will be a pattern approach to the rule, and the second will be a graphical approach to the rule.

Pattern Approach

We will continue the pattern from the previous question:

$$2^4 = 16$$
$$2^3 = 8$$
$$2^2 = 4$$
$$2^1 = 2$$
$$2^0 = 1$$
$$2^{-1} = ?$$

What numbers or values should be placed in the location of the question marks? We will look at the products or values of the expressions to see if there is a pattern. The values of each expression are the following:

$$16, 8, 4, 2, 1, ?$$

The pattern, established in the previous mystery, is that each number is half of the number that precedes it. So the 8 is $16 \div 2$, and the 4 is found by dividing the 8 by 2, and so on.

What number goes where the question mark is? We simply take the number preceding it, in this case the 1, and divide it by 2. The result is $1 \div 2$, or $\frac{1}{2}$.

Therefore, we can see that $2^{-1} = \frac{1}{2}$. If we use other bases, we would find the same results, for example, $3^{-1} = \frac{1}{3}$ and $20^{-1} = \frac{1}{20}$.

A Graphical Approach

If we look at the previous graph and follow it to where the x-axis is negative, we see where the values of negative exponents are fractions. (At least we can see that the graph never crosses the y-axis, or in other words, it never has a negative value.)

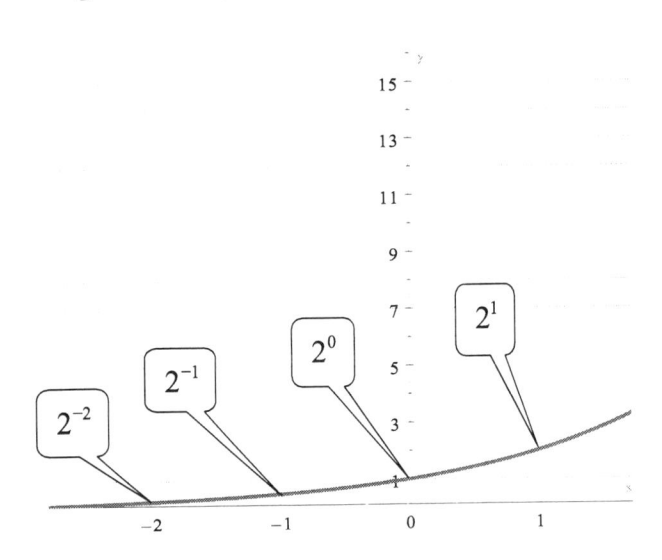

Students tend to be misled by the negative exponent and want to just multiply the base times the negative exponent (getting a negative number).

Showing the graph allows the students to see that in this case, the value of the expression is not negative but equal to a fraction.

To conclude, let's continue the pattern that we started two mysteries ago to show the wonderful symmetry in this pattern:

$$2^4 = 16$$
$$2^3 = 8$$
$$2^2 = 4$$
$$2^1 = 2$$
$$2^0 = 1$$
$$2^{-1} = \frac{1}{2}$$
$$2^{-2} = \frac{1}{2^2} \text{ or } \frac{1}{4}$$
$$2^{-3} = \frac{1}{2^3} \text{ or } \frac{1}{8}$$
$$2^{-4} = \frac{1}{2^4} \text{ or } \frac{1}{16}$$

Questions to Ponder

1. How would you explain why any number raised to the negative one exponent is equal to its reciprocal?
2. How do you remember that when a number raised to the negative one power is equal to its reciprocal?
3. Explain the answer to the expression 5^{-1} without just stating the rule.
4. Write out a real-world context for which you would have to raise a number to the negative exponent.
5. Graph the curve created by raising the number 4 to an exponent. In other words, graph the following:

$$y = 4^x.$$

Mystery #20

Why Do We Do It That Way? Make two equations when Solving Equations or Inequalities with Absolute Values

Short Answer

The absolute value of a number is defined as its distance from zero on the number line. For example, the number 3 is exactly three steps from zero. In fact, there are two numbers three steps from zero. Let's look at the number line.

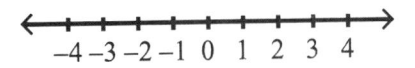

$$-4\ -3\ -2\ -1\ 0\ 1\ 2\ 3\ 4$$

Notice that the number line is symmetric with the number 0 the only number on the number line without a distinct pair or an opposite. This means that for any number on the right-hand side, there exists another number on the opposite side that is the same number of units from zero. For example, the number 3 is three units from zero, but its opposite, −3, is also three units from zero; so for a basic question, such as

$$|x|=3$$

we can essentially have two equations to describe the above relationship.

$$x=3 \text{ and } x=-3$$

With the two equations, we get two answers: 3 and −3. Both numbers are three units from zero.

Detailed Answer

Now we will apply the above explanation to when we have a more complex-looking equation. The same thinking behind an equation as simple as $|x|=3$ can be applied to an equation more complex, such as $|3x|=3$ or $|x+4|=6$.

Most students learn to solve an equation such as

$$|x+4|=6$$

by making two equations. One of the equations must equal the positive, or in this case 6, and the other must equal the negative, or in this case −6. In this case, the two equations are

$$x+4=6 \text{ and } x+4=-6.$$

Then the answer to the original equation is found by solving both equations.

$$\begin{array}{ll} x+4=6 & x+4=-6 \\ \underline{-4 -4} \text{ and } & \underline{-4 -4} \\ x=2 & x=-10 \end{array}$$

Here is the mystery: why do we make two equations?

To understand why we have two equations and consequently two answers, we must simply remember what we discussed earlier: the absolute value is distance from zero on the number line. So it really doesn't matter what is inside the absolute value symbol. Whatever it is will be a certain number of steps from zero. In this scenario, that distance is 6. The question now is why we make two equations with one equaling a positive and the other a negative.

This goes along with what we discussed before. There are two numbers that are 6 steps from zero, 6 and −6. Therefore, to take into account both of these numbers, we set what is inside the absolute value symbols equal to both possible answers.

Therefore, we can take $|x+4|=6$ and write two equations:

$$x+4=6 \text{ and } x+4=-6.$$

Both of which will satisfy the statement that $x+4$ is 6 steps from zero on the number line.

Questions to Ponder

1. How would you explain why two equations are required to solve an equation or inequality with absolute value?
2. Explain how to solve $|3x|=6$. Be sure to give explicit reasons for each step.
3. How do you remember how to solve equations with absolute value, such as $|3x+7|=16$?
4. Explain how to solve $|x+4|<4$. Be sure to give explicit reasons for each step.
5. Explain why you can't have an answer for $|x|<-2$.
6. Explain what the answers for $|x|>-5$ looks like. Be sure to use the definition of absolute value.
7. Solve and graph the answers to the inequality $|x+7|=>2$.
8. Solve and graph the answer to the equation $|x|=5$.
9. Solve and graph the answer to the inequality $|3x+2|\leq 12$.

Mystery #21

What Is PEMDAS?

Short Answer

PEMDAS is the acronym that is commonly used to help students remember the order of operations, namely,

1. **P**arentheses (grouping symbols)
2. **E**xponents
3. **M**ultiplication and **D**ivision (from left to right)
4. **A**ddition and **S**ubtraction (from left to right)

Many teachers have the students remember a sentence where each letter in PEMDAS is the first letter in the sentence. The common sentence is: **P**lease **E**xcuse **M**y **D**ear **A**unt **S**ally.

Detailed Answer

The following is an expression that a class of students might be expected to simplify:

$$24 - 12 \div 3 \times 2 + 8$$

The mystery is, which operation do we execute first? Do we execute $24 - 12$ first? Do we do $12 \div 3$ first? Herein lies the mystery. What is the proper order, and what is the logic behind the order that has been established?

In most classes, one will probably find the order of operations posted somewhere on a poster or a board. The teacher might continuously refer to the poster as the class practices expressions much like the one above. As the students do more and more problems, the teacher becomes more and more confident that the students will remember the order. If the students still don't recall the order efficiently, the teacher will remind them to repeat this sentence to themselves, "Please excuse my dear aunt Sally." If the students remember the sentence, then they will remember the order of operations. (At least this is the hope.)

The problem is that students are a lot like rivers—they will travel the path of least resistance. Students will look at an expression like the one above and do the operation that looks easiest to them. It may even be different from what their friends think is easiest.

The following is a way to help students make some meaning to the order of operations. We must be clear on this point, however—there is no proof as to why certain operations go before another, yet there are a series of questions that one can use to guide the students through to help them connect the order of operations to things other than a simple memorization of a rule or a sentence that has no mathematical background.

The first thing that should be asked when discussing the order of operations is, "Why?" "Why do parentheses come first?" "Why are the exponents next?" And finally, "Why do multiplication and division come before addition and subtraction?" Let the students mull over those questions for a while. In fact, one may decide to do this in a more systematic way by having a discussion about the following situations, with the following question asked before each situation:

If there is a fight, who wins?

$$\times \text{ vs } + \qquad \div \text{ vs } -$$

The students will come up with various reasons for each, such as the following:

- Multiplication is more powerful than addition.
- Division is more powerful than subtraction.

But then you can easily counter with, "Why is multiplication more powerful than addition?" Or "Why is division more powerful than subtraction?" Throughout the various classes that have had this discussion, most come up with the conclusion that multiplication does its operation faster and more efficiently than addition and that division does its operation faster and more efficiently than subtraction.

Now let's look at the next comparisons. If there is a fight, who wins?

$$+ vs \div \qquad - vs \times$$

With $+ vs \div$ and $- vs \times$, the same argument can and usually is followed with the students understanding more or less the same thing as the previous argument.

The fun conversation to have with the students is when you discuss who wins in the following matchups:

$$\div vs \times \qquad - vs +$$

Who wins? Let's take the matchup $\div vs \times$. Many students will say that multiplication wins, but you can counter with, "Why? Why is making things bigger better than making things smaller? Don't they do their operations the same?"

For example, we look at the following two operations on the number 6:

$$6 \times 2 = 12 \text{ and } 6 \div 2 = 3$$

In one instance, the 6 is doubled, and in the other, the 6 is cut in half. Both did basically the same thing to the 6 but in opposite directions.

Eventually, students will come to the realization that multiplication has the same strength as division, and the same goes for the subtraction-and-addition matchup.

Finally, ask the students, "How then do we know which one goes first?" In most cases, a student will come up with "Whoever comes first." But how does one know which one comes first? Eventually, we realize that much like reading, we usually read mathematics from left to right. (Not all the time, but that is a different mystery.)

Don't forget to throw in a parenthesis situation here to discuss when addition or subtraction goes before multiplication or division.

Remember, this is not a proof of the order of operations but a sample dialogue one can have with students to help them build a logical schema to why the operations are in the conventional order.

Questions to Ponder

1. How would you explain the order of operations?
2. How do you remember the order of operations?
3. Some people use a different acronym such as GEMDAS (G = grouping symbols). Why would this one make more sense? And what sentence would you come up with to remember it?
4. Explain how to do the following problem:

$$11 - 8 \times 2 + 6 \div 4 \times 2 - 8 + 4$$

5. How does adding parentheses in the above problem change the value? And why?

$$11 - 8 \times (2 + 6) \div 4 \times 2 - 8 + 4$$

6. Why does an order of operations exist?

Mystery #22

What Is FOIL?

Short Answer

FOIL is an acronym used to help students remember an algorithm for multiplying binomials. For example, to multiply the following two binomials, one would do the following four steps:

1. Multiply the <u>first</u> terms.

   $(x + 2)(x + 4) \longrightarrow x \bullet x = x^2$

2. Multiply the <u>outer</u> terms.

   $(x + 2)(x + 4) \longrightarrow x \bullet 4 = 4x$

3. Multiply the <u>inner</u> terms.

   $(x + 2)(x + 4) \longrightarrow 2 \bullet x = 2x$

4. Multiply the <u>last</u> terms.

   $(x + 2)(x + 4) \longrightarrow 2 \bullet 4 = 8$

If we combine all the products together, we have

$$x^2 + 4x + 2x + 8 \text{ or } x^2 + 6x + 8.$$

In the example above, we multiplied in a specific order. The underlined words give us the order, and the first letters of each of those words compose the acronym FOIL: F = first terms, O = outer terms, I = inner terms, and L = last terms.

Detailed Answer

That is just fine and dandy. But why does it work, and how do we know that the answer that we arrived at is actually the right answer? Let's first look at this from the standpoint that most of us have multiplied like this before and didn't even know it.

For example, let's multiply 25 × 34. The traditional algorithm yields the following product:

$$\begin{array}{r} 25 \\ \times\ 34 \\ \hline 100 \\ 750 \\ \hline 850 \end{array}$$

But let's apply what we learned from a previous explanation. We can write out both the 25 and the 34 into expanded notation, like this:

$$25 = 20+5$$
$$34 = 30+4$$

Now we can set up the multiplication as follows and apply FOIL to see what the answer would be.

$$(20+5)(30+4)$$

1. Multiply the <u>first</u> terms.

$$(20+5)(30+4) \longrightarrow 20 \bullet 30 = 600$$

2. Multiply the **outer** terms.

$$(20+5)(30+4) \longrightarrow 20 \bullet 4 = 80$$

3. Multiply the **inner** terms.

$$(20+5)(30+4) \longrightarrow 5 \bullet 30 = 150$$

4. Multiply the **last** terms.

$$(20+5)(30+4) \longrightarrow 5 \bullet 4 = 20$$

If we combine all the products together, we have

$$600 + 80 + 150 + 20, \text{ or } 850,$$

It works! Also, make note that it would not have worked out had we not multiplied the 20 by both the 30 and the 4 and the 5 by both the 30 and the 4.

So what is really going on? Well, we are using the distributive property. Just as stated above in the problem, $(20+5)(30+4)$, we had to multiply both the 20 and the 5 by the 30 and the 4. If a student simply remembers the distributive property when multiplying polynomials, then the acronym FOIL is not needed.

For example, we can look at the previous example again, but state it this way:

$(20+5)(30+4)$ Distribute the 20 to both the 30 and the 4.
$20 \bullet 30 = 600$ and $20 \bullet 4 = 800$

$(20+5)(30+4)$ Distribute the 5 to both the 30 and the 4.
$5 \bullet 30 = 150$ and $5 \bullet 4 = 20$

If we then combine the products, we get $600 + 80 + 150 + 20$, or 850. As we can see, we arrive at the same answer, but discussing the technique from the context of the distributive property allows students to see and understand the mathematics behind FOIL.

In fact, if the distributive property is the center of attention when discussing these problems, then a problem such as

$$(x + 2)(x^2 + 3x + 2)$$

will not cause students much trouble.

Questions to Ponder

1. How would you explain the process of multiplying two binomials or two polynomials?
2. How do you remember the steps?
3. Explain how to multiply: $(x + 8)(x - 6)$.
4. Why do you think teachers teach the FOIL method?
5. Is it beneficial to teach the FOIL method?
6. When would it be beneficial to teach the FOIL method?
7. Would multiplying two binomials yield the same result no matter which order it was done in? For example, instead of FOIL, do OLIF, LOIF, or IOLF.

Mystery #23

What Does It Mean to "Complete the Square"?

Short Answer

It is the process of manipulating a polynomial with the desired result being a perfect square trinomial. Completing the square is used to solve quadratic equations and manipulating the different forms of a conic section (see glossary: *conic section*).

Detailed Answer

Completing the square comes up when manipulating polynomials while solving equations or simplifying equations of conic sections, such as circles and parabolas. For example, the following polynomial may be present in an equation to solve

$$(x^2 + 6x).$$

The steps taught in a typical high school math class may include the following:

1. First, take the coefficient in front of the *x* term and divide it by 2. (In this case, we take 6 and divide it by 2, or $6 \div 2 = 3$.)
2. Second, we take that number (in this case 3) and square it ($3^2 = 9$).
3. Now take this number and add it to the polynomial. (If solving an equation, you would add this number to both sides of the equation.)

The resulting polynomial is

$$(x^2 + 6x + 9).$$

You have now completed the square!

Students, hopefully, will now ask, "Why?" Or at least they will ask the teacher to explain the mathematics behind the steps used. For example, why do we divide the 6 in half, or why do we square the 3?

We will show why we do these traditional steps through the use of algebra tiles.

Algebra tiles are used to help visualize many concepts in mathematics, especially in algebra, hence the name. There are three different sized blocks:

⟶ x^2 – Think of this block as a square with side lengths of x

⟶ x – This block has one side of x and the other of 1.

⟶ 1 – The unit block. This block is a square with side lengths of 1.

Now we will represent the polynomial $x^2 + 6x$ with the algebra tiles.

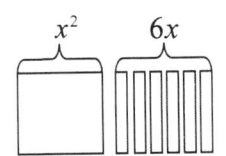

Notice that we have one x^2 and six x tiles. If we want to illustrate completing the square, we will want to try and manipulate the tiles around until we have a perfect square. (Remember that a square has sides of the same length.) After trying several different configurations of the tiles, the following is the best and closest configuration of a square:

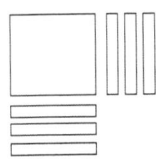

Notice that this is not a complete square. To complete the square, we must add some units to it.

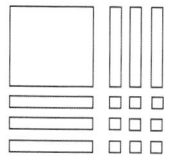

We have now added enough unit blocks to complete the square. This square has side lengths of $x + 3$.

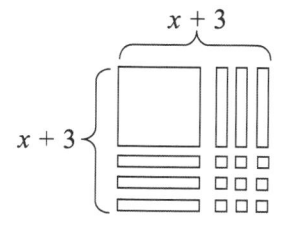

$x + 3$

$x + 3$

But what did we just do? Let's first look at the incomplete square.

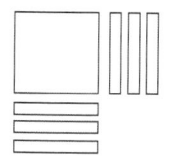

Notice that the x tiles have three lined up along the right-hand side of the x^2 tile and three lined up along the bottom of the x^2 tile. Essentially, we have taken the six x tiles and split them into two equal groups of three. (Remember when we divided the 6 by 2?)

Now let's look at the completed square with the tiles that we added.

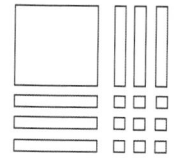

We added exactly 9 tiles to complete the square. We should also point out that these 9 tiles are configured into their own little square as well. The side lengths are especially noteworthy. This square has side lengths of 3. Remember when we squared the 3 to get the 9 in the beginning?

We can now see where the steps come from—the dividing the coefficient of the x term and the squaring of that number to get what needs to be added to complete the square.

Notes

We did not cover what the square looks like when there is a coefficient in front of the x^2 term. The same approach can be taken to show this as well, but this goes beyond the scope of this book. Perhaps in future editions we will explore these, but for now, have fun exploring on your own!

Questions to Ponder

1. How would you explain what it means to complete the square?
2. How do you remember the steps?
3. How would you complete the square for a polynomial such as $x^2 - 6x$? Make a diagram.
4. How would you complete the square for a polynomial such as $2x^2 + 8x$? Make a diagram.
5. Does this process work with numbers? For example, could you show how to complete the square for a number such as 11? What would this look like? Could you describe it with the notation used in this chapter?

Mystery #24

Why Does 0! = 1?

Short Answer

If we look at factorials for what they are (ways in which to arrange objects), we can answer this intuitively.

Example:

> 4! = How many ways can you arrange 4 objects?
> *The answer is 24, or $4 \times 3 \times 2 \times 1$.*
>
> 3! = How many ways can you arrange 3 objects?
> *The answer is 6, or $3 \times 2 \times 1$.*
>
> 2! = How many ways can you arrange 2 objects?
> *The answer is 2, or 2×1.*
>
> 1! = How many ways can you arrange 1 object?
> *The answer is 1, or 1×1.*
>
> 0! = How many ways can you arrange 0 objects?
> *The answer is 1.*

It may take you a second to see why the number of ways to arrange zero objects is only one, but there you go.

Detailed Answer

In probability and statistics, there is something called a permutation. A permutation is an ordering of a number of objects. For example, there are 6 permutations of the letters A, B, and C.

ABC   ACB   BAC   BCA   CBA   CAB

If one would want to find out the number of permutations without writing them all out, one would only need to use the fundamental counting principle. The fundamental counting principle states that if there are m ways to do one thing, n ways to do another thing, p ways to do a third thing, and so on, then the number of ways of doing all those things at once is $m \cdot n \cdot p$. For example, if we go back to the letters A, B, and C, there are 3 choices for the first letter. After the first letter has been chosen, 2 choices remain for the second letter. Finally, after the first two letters have been chosen, there is only 1 choice remaining for the final letter. So the number of permutations is $3 \cdot 2 \cdot 1 = 6$.

The expression $3 \cdot 2 \cdot 1$ can be written as 3! The symbol ! is the factorial symbol, and 3! is read as "3 factorial." In general, $n!$ is defined where n is a positive integer as follows:

$$n! = n \cdot (n-1) \cdot (n-2) \cdot \ldots 3 \cdot 2 \cdot 1$$

Trying to find a pattern with factorials provides the following problem or mystery.

Observe:

$4! = 4 \cdot 3 \cdot 2 \cdot 1 = 24$

$3! = 3 \cdot 2 \cdot 1 = 6$

$2! = 2 \cdot 1 = 2$

$1! = 1 \cdot 1 = 1$

$0! = ? = 1$

> Herein lies the mystery. What replaces the ? in order to make the statement true?

Below is another way to look at the above set of factorials. It must be made clear that this is not a proof either but an illustration of the definition of 0!

A Pattern Approach

Let's look at the factorials that we listed before again.

$$4! = 4 \cdot 3 \cdot 2 \cdot 1 = 24$$

$$3! = 3 \cdot 2 \cdot 1 = 6$$

$$2! = 2 \cdot 1 = 2$$

$$1! = 1$$

$$0! = ?$$

What goes where the question mark is? If we look at the products above, we might discover a pattern. The products are as follows:

$$24, 6, 2, 1, ?$$

There is a pattern in these numbers; each number is being divided by numbers in succession by one less than the previous set. In other words, to get the 6, we divide 24 by 4, and to get the 2 we divide the 6 by 3, and to get the 1 we divide the 2 by 2.

If we follow this pattern, 4, 3, 2 . . . , the next number we would be dividing by is a 1. Therefore, if we take the 1 and divide it by 1, we get the value that goes where the question mark is, which is 1. Therefore, 0! = 1.

Another Approach

Let's try and show it this way:

$$4! = \frac{5!}{5} = \frac{5 \cdot 4 \cdot 3 \cdot 2 \cdot 1}{5} = 24$$

The 5s are canceled out, which yields $4 \cdot 3 \cdot 2 \cdot 1$, which is 4! To the left, the pattern is continued until 0! is defined.

$$3! = \frac{4!}{4} = \frac{4 \cdot 3 \cdot 2 \cdot 1}{4} = 6$$

$$2! = \frac{3!}{3} = \frac{3 \cdot 2 \cdot 1}{3} = 2$$

$$1! = \frac{2!}{2} = \frac{2 \cdot 1}{2} = 1$$

$$0! = \frac{1!}{1} = \frac{1}{1} = 1$$

Following the pattern set above allows us to visually account for the basic definition of 0! and this gives us another logical reason for why it is defined this way.

Unfortunately, this does not prove the definition of 0! But this is an approach that should satisfy many and give at least most students a reason for the definition.

Questions to Ponder

1. How would you explain the reason that 0! = 1?
2. How do you remember that 0! = 1?
3. Why do factorials exist?
4. What would happen if $0! \neq 1$?
5. Why do you think the exclamation point was used for factorials?

Mystery #25

What Is π?

Short Answer

π, spelled *pi* and pronounced like the dessert, "pie," is one of the most intriguing numbers in the world. It is also probably the most popular number in the world. It has quite a fan base.

π describes the ratio of the circumference of any circle to its diameter. As a formula it looks like this:

$$\pi = \frac{C}{d}$$

The ratio is usually written in decimal form as 3.14 . . .

The reason the Greek letter π was chosen to represent this number goes back to the word in Greek meaning perimeter, which is περιμετρος. The symbol π is the first letter in this word, and it would be a logical assumption to infer that this is where the use of this Greek letter originated.

Detailed Answer

Here are a few more explanations and trivia about this most useful number:

- π is the symbol for the Greek letter *pi*.
- π is a nonrepeating, nonterminating decimal.
- π is a number used to find the area and circumference of a circle.
- π is used in statistics and probability and in other academic branches.

All the above explanations and more are used by teachers everywhere to describe arguably the world's most famous number, except for perhaps the numbers 1 and 0.

But do these descriptions adequately describe what π is?

The following are some simple explanations that can help students visualize what pi (π) is and what it is used for in mathematics.

An Object Lesson Approach

One engaging way of illustrating the definition of π to students is by showing them a demonstration. The object is to show that the circumference is a little over three times the diameter or 3.14159...

Do this by getting some string and wrapping it around something circular, like the base of a round trash can. The string is the circumference. Do this as exact as possible. Now take the circumference (the string) and stretch it over the diameter of the base of the trash can. You will be able to drag the string across the base of the trash can three times (3.00) and there will be a little bit left over (0.14159...).

Algebraic Approach

Another way of showing what π is to the students is by taking the basic circumference formula of a circle and solving for π. This shows the students the relationship of π to the circumference and diameter of a circle.

$$C = \pi d$$
$$\frac{C}{d} = \frac{\pi \cancel{d}}{\cancel{d}}$$
$$\frac{C}{d} = \pi$$

The above formula manipulation is a quick and simple way of answering the students' question, "What is π?"

The key to helping students understand that π is not a mystery is to have an explanation for them when the question arises. It does not have to be elaborate or complex, but a teacher does need to have one, or the students will not see π as equaling 3.14159 ... but just another thing that they have to memorize for a test.

Questions to Ponder

1. How would you explain what π is?
2. How do you remember the decimal equivalent to π?
3. Do you know of any other ways of explaining π? If so, what are they?
4. Many people celebrate Pi Day, March 14, what types of activities could be used to celebrate this day?
5. Are there any other days during the year that could be a "Mathematical Celebration?" For example, September 10, 2011 could be called "Counting Day."

Mystery #26

What Is e?

Short Answer

Just like the number π, e is a number whose value is a nonrepeating, nonterminating decimal, or in other words, it is an irrational number approximated by the decimal 2.71828 . . .

Detailed Answer

Much like π, e has been known as a mystery to students for a long time. More so than π, however, is the fact that it is usually more of a mystery to teachers than π, meaning teachers in general will not have a firm foundation of what this number represents beyond that of its decimal equivalent.

The most common definition for e is that the function e^x is equal to the slope of the tangent line for any value of x. In other words, the steepness of the line at any value of x is equal to the value of x. Below is an illustration of e^x (the curve) and the tangent line at 1.

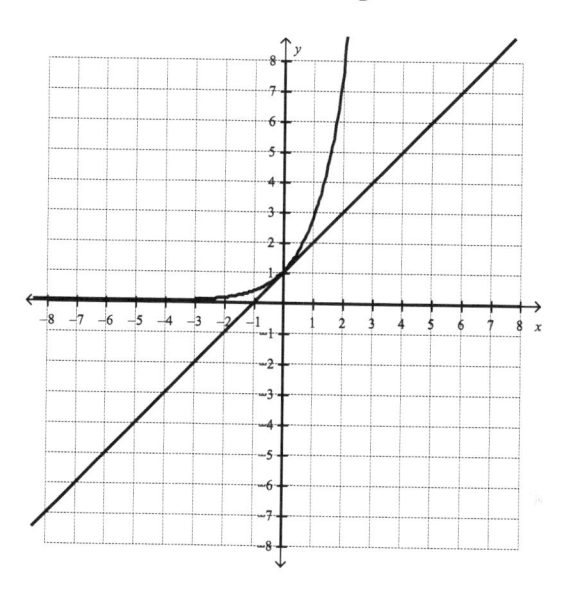

We will first go over other simple definitions of e.

e is found in many mathematical formulas that describe a nonlinear increase or decrease such as growth, decay, compound interest, the statistical bell curve, the shape of a hanging cable, or a standing arch.

e is found in some problems of probability, some counting problems, and even in the study of the distribution of prime numbers.

e occurs naturally with some frequency in the world. For example, it is the base for the natural logarithm.

e can also be defined by the following equation:

$$e = \lim_{n \to \infty} \left(1 + \frac{1}{n}\right)^n$$

What Is e Used For?

- It is the base for the natural log (John Napier).
- It is used for the study of prime numbers.

- It is used to help calculate the compound interest in economics.
- It is used in architecture.
- It is used in physics.
- It is used in geology to calculate the decay of radioactive material.

Who Developed or Discovered the Number e?

The number *e* was first studied by the Swiss mathematician Leonhard Euler in the 1720s. Although an earlier mathematician, John Napier, implied that such a number existed in 1614, Euler was the first to use the letter *e* for it in 1727. The number *e* is sometimes referred to as the Euler number although this tends to be confused with another one of Euler's concepts that deals with geometry.

Disclosure

As you can see, even this explanation doesn't satisfy my need to make it simple to understand. *e* is very much a mystery until one reaches higher mathematics, but by then, students usually take *e* as a constant, such as π, and just work out the problems.

However, I believe it is beneficial at all levels to revert to basic definitions when discussing any topic. So it is with *e*, a number equaling 2.71828 . . . and for which we may need to search further for a simple and elegant definition.

Questions to Ponder

1. How would you explain *e*?
2. What would a visual of *e* look like?
3. How do you remember the decimal equivalent for *e*?
4. Many see the following formula one of the most interesting. Why do you think that is so?

$$e^{\pi i} + 1 = 0$$

5. Research other ways in which *e* is used in real-world contexts.

Mystery #27

What Is i?

Short Answer

The number i is equal to $\sqrt{-1}$. In other words, $i = \sqrt{-1}$.

Detailed Answer

The number i is imaginary. It is imaginary because it satisfies the simple question, "What happens if you take the square root of a negative number?"

Why is taking the square root of a negative number so imaginary? Let's discuss first the concept of taking the square root of a number.

Taking the Square Root of a Positive Number

When taking the square root of a number, one finds the value that when multiplied by itself will result in the original number.

For example, let's take the number 36. If we want to take the square root of 36, written as $\sqrt{36}$, we would find the number that when multiplied by itself will result in 36. There are actually two possible values for this 6 or -6.

In other words, 6 • 6 = 36 and (−6)(−6) = 36.

Important note: $\sqrt{36}$ usually implies only the positive roots, but we showed both the positive and negative roots here for a reason that will hopefully become apparent soon.

Taking the Square Root of a Negative Number

When taking the square root of a negative number, the same situation must be satisfied—what value multiplied by itself will result in the desired number? Let's look at the number −36.

If we take the square root of −36 or $\sqrt{-36}$, we must be able to find a number that when multiplied by itself will equal −36. This is impossible because we already know that 6 • 6 = 36 and (−6)(−6) = 36; to get −36, we would need one of the numbers to be negative and the other to be positive.

Remember that 6 × (−6) = −36, but it does not satisfy $\sqrt{-36}$ because 6 and (−6) are not the same number.

To be able to take the square root of a number, the imaginary number *i* was developed. It is imaginary because, intuitively, the square root of a negative number should not exist. The basic definition of *i* is found below:

$$i = \sqrt{-1}$$

Therefore, returning to our previous question, $\sqrt{-36}$, we could break up −36 as (−1)× 36, then we would have

$$\sqrt{(-1) \cdot (36)}.$$

And then taking the square root of each number separately, we would have

$$\sqrt{(-1)} = i$$

and

$$\sqrt{36} = 6.$$

the answer would then be 6*i*. The *i* in the result represents the square root of the negative one.

Who Developed the Number i?

Rene Descartes, who coined the phrase "I think therefore I am," was the first to use the term *imaginary* when describing $\sqrt{-1}$ in 1637. However, the concept of imaginary numbers was discovered much earlier by Gerolamo Cardano in the 1500s.

Questions to Ponder

1. How would you define i?
2. How do you remember that i equals $\sqrt{-1}$?
3. Explain how to take the square root of (-49).
4. Answer question four from the previous mystery again.
5. Make a table for i raised to several consecutive powers. What pattern can you see?

Mystery #28

Why Are There 360° in a Circle?

Short Answer

Divide a circle into 360 parts. One degree (d) is defined as the angle formed by the line segments that create one of those parts, in this case m.

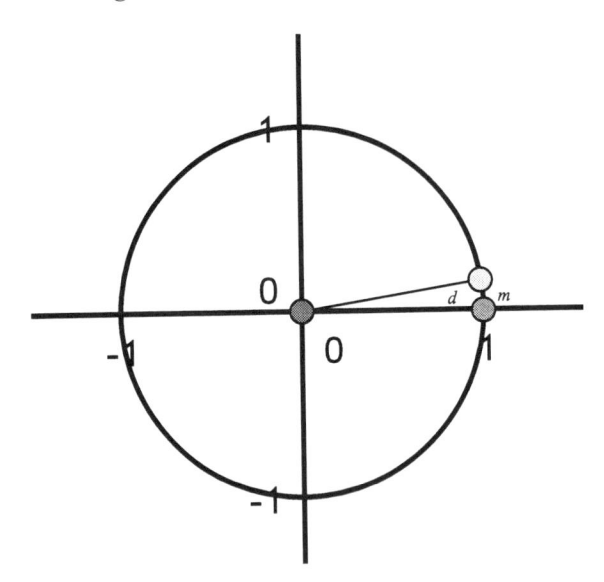

The distance between the two points, m, along the outside of the circle is exactly $\frac{1}{360}$, the length of the entire circumference of the circle. In other words, if we had 360 of these, we would have an entire circle.

It is important to note as well that the arc length (m) and the angle measure (d) are equal.

Detailed Answer

Degree is the most common way of measuring the angle between two lines. The real question here, though, is, why was the circle divided up into 360 parts and not some other number, like 100? After all, 100 is perhaps easier to divide and much smaller than 360. Wouldn't it have been easier to use a smaller number?

The reason for using 360 is partly due to the fact that it is very composite, meaning it has a lot of factors. There are 24 numbers that divide 360. It appears that the Greeks popularized the 360 degrees in a circle based on some influence from the ancient Babylonians. The Babylonians had a number system based on 60, and they not only divided the circle into 360 equal portions but also used a 360-day calendar. We still see their influence today: we divide the hour into 60 minutes and a minute into 60 seconds.

The number of factors found in 360 allows the circle to be easily divided into various parts. One of the more popular ones is that of the 90-degree angle, which is exactly a quarter of the full circle, or the 60-degree angle. This angle is found in equilateral triangles.

So I guess to make a long story short, we use 360 to describe the number of degrees around a circle because it is a very useful way to divide up the circle, and this usefulness helps in other branches of mathematics.

Questions to Ponder

1. How would you explain why there are 360 degrees in a circle?
2. What would happen if 100 was used to divide up a circle? For example, how many degrees would be a fourth the way around? (Hint: for our current system, 90 degrees is a fourth the way around the circle.)
3. Find the equivalencies in a 360-degree circle and a 100-degree circle:
   90° = _____
   45° = _____
   60° = _____
   30° = _____
   180° = _____
   100° = _____
   360° = _____

Mystery #29

What Is a Radian?

Short Answer

A radian is an angle measure used in higher math courses such as trigonometry and calculus. To be specific, a radian is the measure of a central angle that intercepts an arc equal in length to the radius of the circle. Below is a simple illustration.

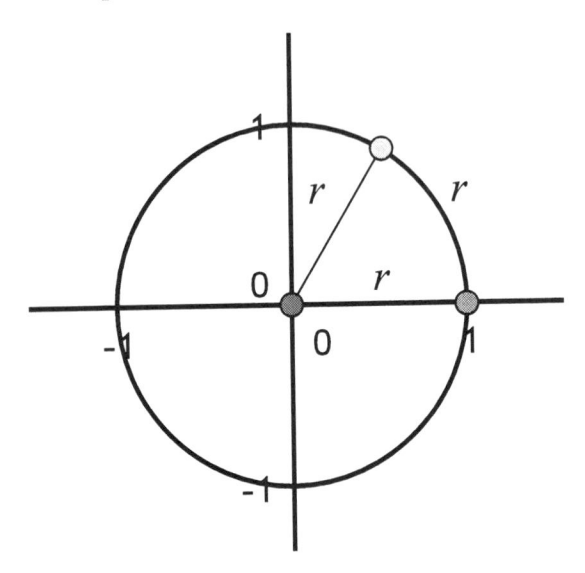

In the case above, the radius is equal to 1 and so is the length of the arc (meaning the distance from one point to another along the circle) that is formed by the two radii above. I have labeled all three of them *r* because of this relationship.

Detailed Answer

Since the above explanation satisfies the purpose of this book and is extensive enough to qualify for a detailed answer as well, we will use this section to compare a radian with a degree.

It is very simple really. A degree describes the relationship between the angle measure and the arc length formed by the angle. In other words, the angle measure is equal to the arc length. A radian describes the relationship between the length of the radius and the arc length. The radius, in this case, is equal to the arc length, not the angle.

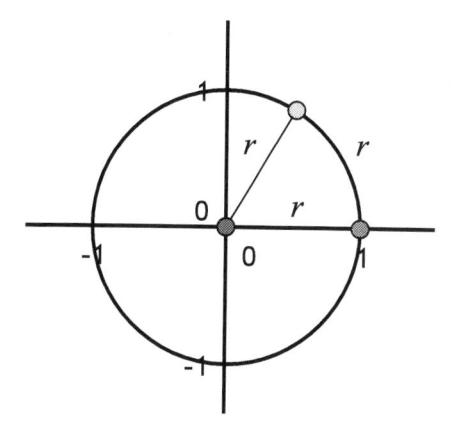

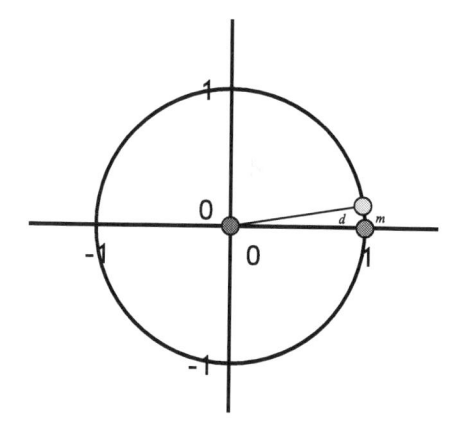

A radian is the relationship between the radius and the arc length equal in length to the radius. In this case, the radius and the arc length are equal.

A degree is the relationship between the angle and its corresponding arc length. In this case, the angle and the arc length are equal.

Classroom Insight

If one were interested in doing a demonstration of a radian, one could get the following materials:

- Something round and clear, like a magnifying glass or a ring. (You don't want it too small though.) This will be your circle.
- A piece of string.
- A black marker.
- A piece of graph paper.

Here are the steps:

- Draw a Cartesian graph on the piece of graph paper.
- Take the circle and place it directly on top of the Cartesian graph. You want to make sure that the center of the circle is at the origin.
- Trace the circle. This is your unit circle. From the origin of the circle to the intersection of the axes should be labeled as one unit.
- Take the string and lay it across the circle with one end at the origin and the other end through the circle.
- Take the black marker and mark on the string where the string and the circle intersect.
- You may decide to continue this procedure until you have made 6 other marks of the same length on the string.
- Pick up the string and lay the end that was at the origin where the positive x-axis intersects the circle.
- Lay the string onto the circle, and where the black marks touch the circle, make a mark.
- From this point, draw a line to the origin. You have now constructed a radian.
- If you made the other marks on the string, you may want to go ahead and divide up the circle into radians. We will use it for the next mystery.

Questions to Ponder

1. How would you explain what a radian is?
2. How many degrees are in one radian? How would you figure it out?
3. Besides the degree and radian, is there another way to divide up a circle?
4. Why was the name radian chosen, do you think, to describe the method of partitioning a circle?
5. Which one is more useful, the degree or the radian?

Mystery #30

Why Are There 2π Radians in a Circle?

Short Answer

Why are there 2π radians in a circle? Because the circumference of a circle is $C = 2\pi r$. Remember that the length of the arc of one radian is equal to the radius, r. Therefore, there are, according to the circumference formula, 2π radius lengths around a circle. This also suggests that there are 2π radians around a circle as well.

Detailed Answer

Remember that the radian describes the relationship between the radius and the arc length; they are equal. In the previous mystery, we divided up the circle into only one radian as shown below.

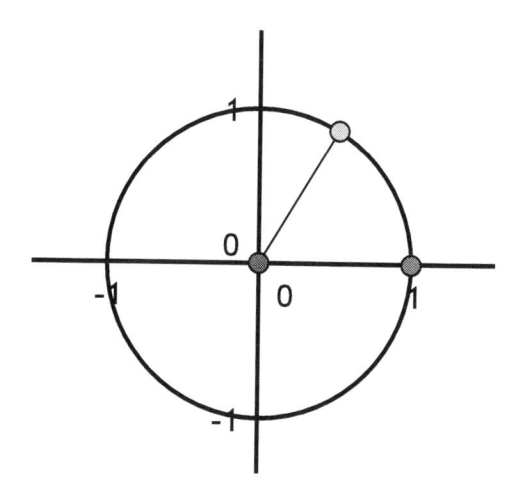

We will now divide the rest of the circle up into radians and see what we can discover.

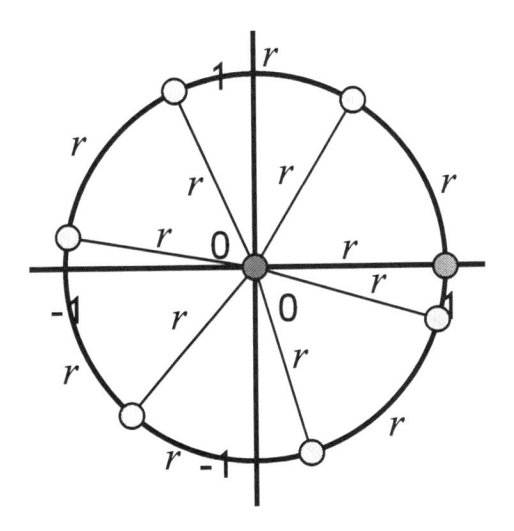

There are a lot of *r*'s in this diagram. To me, that is part of the beauty of it. Each one of the lines or arcs next to an *r* is the same length, in this case 1.

Let's look at the diagram a little closer but take the *r*'s away.

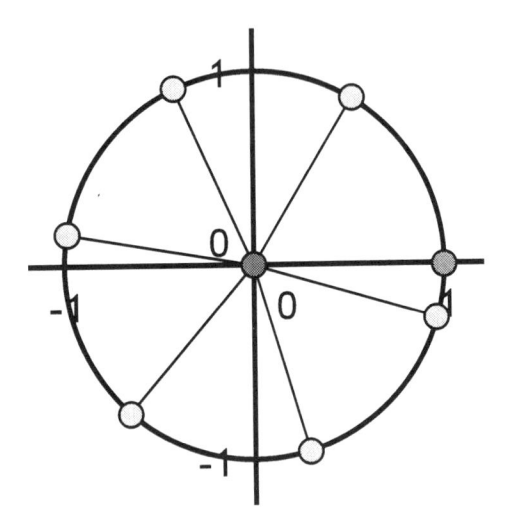

I want you to look at the diagram a little closer now and notice something. How many full radians are there halfway around the circle?

Three. Notice that the third point moving counterclockwise does not quite reach the *x*-axis again. I will point this out below.

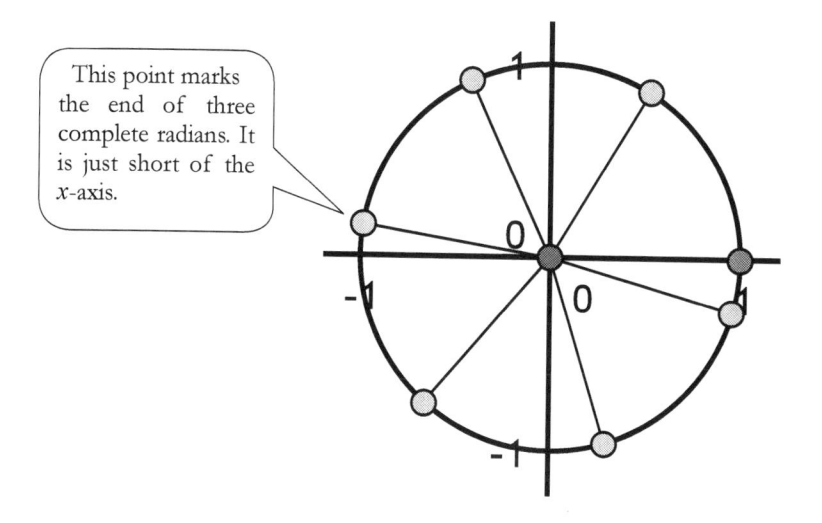

This point marks the end of three complete radians. It is just short of the *x*-axis.

How significant is that? The significance is on how much it is short. Without going through too much proof, it is exactly 0.14159 . . . (radians) short of the *x*-axis. I know what you are thinking! If we combine that number with the three full radians already drawn in this semicircle, we have 3.14159 . . . radians that cover half of the circle. In other words, there are π radians in a semicircle.

What implication does this have on the entire circle? Well, if we double this length, π, we get 2π, which is the number of radians in a circle.

Classroom Insight

If one were interested in doing a demonstration of how many radians are in a circle, one could get the following materials:

- Something round and clear, like a magnifying glass or a ring. (You don't want it too small though.) This will be your circle.
- A piece of string.
- A black marker.
- A piece of graph paper.

Here are the steps:

- Draw a Cartesian graph on the piece of graph paper.
- Take the circle and place it directly on top of the Cartesian graph. You want to make sure that the center of the circle is at the origin.
- Trace the circle. This is your unit circle. From the origin of the circle to the intersection of the axes should be labeled as one unit.
- Take the string and lay it across the circle with one end at the origin and the other end through the circle.
- Take the black marker and mark on the string where the string and the circle intersect.
- Repeat this process with the same piece of string, moving each dot to the origin to make each succeeding mark on the string. (You will want to make at least six marks.)
- Pick up the string and lay the end that was at the origin where the positive x-axis intersects the circle.
- Lay the string onto the circle, and where the black marks touch the circle, make a mark.
- From each point, draw a line to the origin. You have now divided up the circle into radians.
- Have students label the radii, the arc lengths, and point out some of the important intersection points of the circle. A few of them are labeled below.

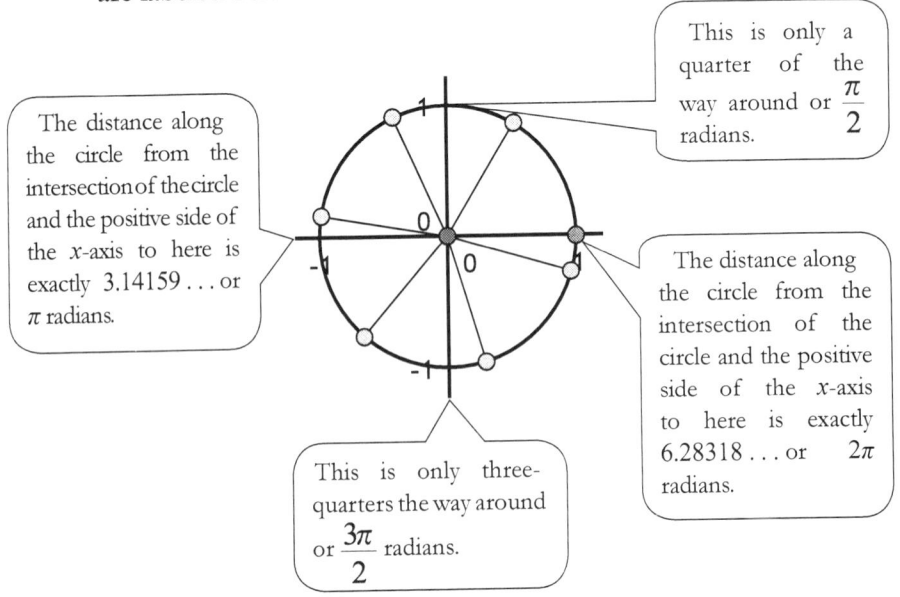

This is only a quarter of the way around or $\dfrac{\pi}{2}$ radians.

The distance along the circle from the intersection of the circle and the positive side of the x-axis to here is exactly 3.14159... or π radians.

The distance along the circle from the intersection of the circle and the positive side of the x-axis to here is exactly 6.28318... or 2π radians.

This is only three-quarters the way around or $\dfrac{3\pi}{2}$ radians.

Questions to Ponder

1. How would you explain why there are 2π radians in a circle?
2. Why do you suppose π is present in a radian?
3. Where do you think the word *radian* comes from?
4. Some have suggested that 2π is a more important or useful number than π. Why do you think that is so?
5. Compare 2π and $360°$.

Conclusion

The essence of mathematics is not to make simple things complicated, but to make complicated things simple.
—S. Gudder

Many teachers and students see mathematics as a list of discrete procedures and definitions that need to be memorized. Teachers list off definitions, work out examples on the board, have the students mimic the steps on similar problems, and then, if the students can duplicate the steps on several problems, they have "mastered" it.

Mathematics is much more than procedures and steps to memorize. It is the quantification of the world around us, pattern finding, data analysis, exploration of ideas, reasoning, and logic. It is not mysterious. Every procedure, rule, definition, step, and operation has a purpose in mathematics and a clear reason behind it. If teachers and students examine the reasoning behind a procedure or technique, they will gain an increased comprehension of mathematics and understand that mathematics will make sense no matter at what level they are learning or teaching.

Mathematics is not truly mastered until a reason can be given for the procedure, a logical explanation can be derived for the operation, and a purpose can be placed upon the step. Mathematics taught with this in mind enables both the student and the teacher to grasp a deeper meaning of mathematical concepts. Most importantly, mathematics ceases to be the mystery that it commonly gets labeled. The mystery is not in mathematics but in how we teach and learn mathematics.

There is a world of difference between the student who can summon a mnemonic device such as "FOIL" to expand a product such as $(a + b)(x + y)$ and a student who can explain where the mnemonic comes from. Teachers often observe the difference firsthand, even if large-scale assessments in the year 2010 often do not. The student who can explain the rule understands the mathematics, and may have a better chance to succeed at a less familiar task such as expanding $(a + b + c)(x + y)$. Mathematical understanding and procedural skill are equally important, and both are assessable using mathematical tasks of sufficient richness. (Draft of the CCSS)

Finally, we quote Roger Bacon, who once said, "Mathematics is the gate and key of the sciences . . . Neglect of mathematics works injury to all knowledge, since he who is ignorant of it cannot know the other sciences or the things of this world. And what is worse, men who are thus ignorant are unable to perceive their own ignorance and so do not seek a remedy" (*Opus Majus*, 1900, book 1, chapter 4).

So I hope that you have enjoyed this little journey through some of the mysteries of mathematics. My hope was that when you finished reading this book, you would see mathematics in a different way.

Maybe you read an explanation and said to yourself, "If my teachers had told it to me that way, I would have done so much better in math" or "I wish my teachers had explained math that way to me. I might have actually liked math."

Maybe you read an explanation and caught a new insight into something that you already understood previously and you now understand it a little better.

Hopefully, everyone comes away from this work with a greater appreciation for mathematics. And everyone can see better that all the concepts and all the definitions and all the rules are related somehow.

Finally, I hope you can see that I have great respect for this subject. I appreciate that it can all make sense and that it makes sense of the world around us, even when humans don't make sense. Mathematics is a wonderful subject, and I am honored to be associated with it.

GLOSSARY

The following list is of phrases, words, and concepts that are referenced throughout the book. They are very brief definitions and illustrations with the purpose of helping readers to understand the explanations in the book. They are not technical or exact but written with the purpose of providing a general background.

A

absolute value	*Literally speaking, it signifies the distance of a number from zero on the number line.*
algorithm	*A set of procedures or steps to arrive at a desired result. An example of an algorithm is the traditional formatting and technique used to do long division.*
angle	*A corner of a shape, such as a square or a rectangle. It can also be drawn with two lines that intersect.*

B

base ten blocks	*Instructional materials that are used to illustrate the traditional place value system (base ten). There are generally three types of blocks—the ones, the tens, and the hundreds—although one may also find a thousands block as well.*
binomial	*A polynomial with only two terms. For example, $x + 2$ is a binomial because it only has two terms, x and 2.*

C

Cartesian graph

Consists of two axes, the x-axis and the y-axis. This tool is used to analyze the relationship between values in the form of lines, points, shapes, and curves. An example of a blank one is shown below.

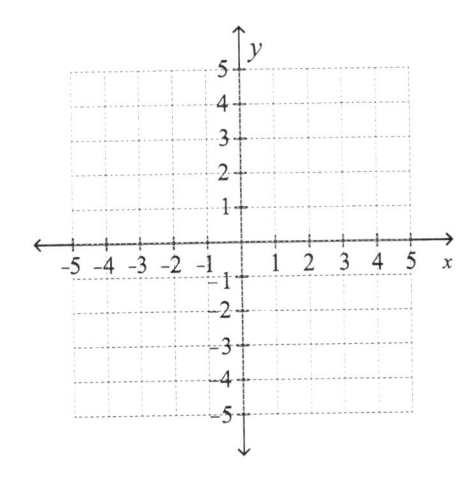

circle

A closed curve where every point on the curve is the same distance from a point called the center of the circle.

conic section

If a cone is taken and sliced in a certain way, the resulting cut portion is a conic section. There are several types of conic sections. A few of them are the circle, the hyperbola, the ellipse, and the parabola. A diagram of these is given below.

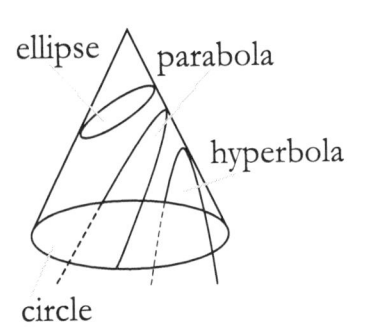

cosine	One of the six fundamental trigonometric functions. The cosine of a right triangle is usually described as a ratio, or fraction. The ratio is of the adjacent side of a triangle over that of the hypotenuse.

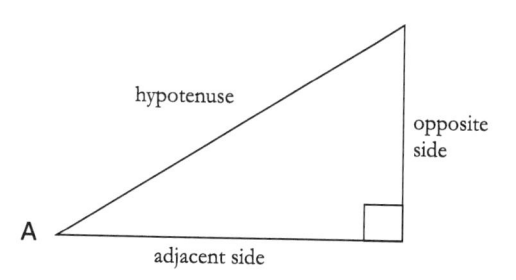

[Diagram of a right triangle with vertex A on the left, hypotenuse along the top, opposite side on the right, and adjacent side along the bottom.]

The diagram above is of a right triangle. If we were to find the cosine ratio, we would put the length of the adjacent side over the length of the hypotenuse.

D

decimal	When this word is used, it usually refers to the "." (point) in a number such as in the number 14.98 and is called the decimal point. However, it also refers to the number system that is commonly used throughout the world, the decimal system or base ten system.

denominator	In a fraction, this refers to the number or value located at the bottom portion. For example, the number 5 is the denominator in the fraction

$$\frac{3}{5}.$$

difference	This usually refers to the distance between two numbers on the number line or the answer when a subtraction operation is performed. For example, the difference between the number 7 and 3 is found by doing the operation $7 - 6 = 4$ or by looking at the distance between the numbers 7 and 3 on the number line.

dividend — *The value that is being divided up into equal parts by a divisor. For example, the number 12 is the dividend in the following problem, written in the standard algorithmic form:*

$$5\overline{)12}$$

divisor — *The value that is doing the dividing of a number or dividend. For example, the number 5 is the divisor in the following example, written in the standard algorithmic form:*

$$5\overline{)12}$$

E

equation — *A mathematical statement that states that two expressions have the same value. For example, in the following equation,*

$$3x + 4 = 2x - 5,$$

the two expressions $3x + 4$ and $2x - 5$ are stated here as having equal value.

expanded notation — *A way to write out numbers that exposes the place value of each digit in a number. For example, the number 254 can be written as*

$$200 + 50 + 4.$$

This notation reveals the true value of the digits in the number 254. The value of the 2, for example, is not 2 but 200.

exponent — *The small number superscripted to the right of a number that indicates the number of times a number is multiplied to itself. For example, 4^3 means*

$$4 \times 4 \times 4.$$

F

factorial

Notation used for computing the number of arrangements possible for a certain number of objects. For example, if you have 4 objects (distinct), then there are 4! ways of arranging those objects, or

$$4! = 4 \times 3 \times 2 \times 1 = 24.$$

fraction

A number or value on the number line that describes some of the numbers between the integers. Some of the other numbers between the integers are the irrational numbers.

G

graph

This word refers to two things: (1) the actual Cartesian graph and (2) the process of plotting points, drawing points that satisfy an equation, connecting those points to draw a line, curve, or other object.

I

improper fraction

A fraction where the numerator is greater than the denominator. Example:

$$\frac{12}{7} \text{ or } \frac{5}{4}$$

inequality

An inequality is a way to compare two or more things, usually to denote that one thing is greater than (>) or less than (<) the other.

infinitesimal

A number so small that it cannot be distinguished from zero. For example, on the number line, an infinitesimal is the value as close to zero imaginable without actually being zero.

integer	*Integer is the name given to all the counting numbers and their opposites, the negatives. In other words, the integers are all the whole numbers on both sides of the number line, positive and negative.*
invert	*The act of rewriting a number (or expression) as its reciprocal.*
irrational number	*A number that cannot be represented exactly as a fraction. For example, one common approximation to π is the fraction $\frac{22}{7}$. But it is not the exact value. The exact value is the nonrepeating decimal that goes on forever.*

L

long division	*The standard algorithm used for dividing numbers.*

M

manipulative	*Tangible objects used in mathematics to help students visualize math concepts. Some examples are base ten blocks, algebra tiles, prime number blocks, pattern blocks, fraction tiles, and unifix cubes.*
mixed number	*A number with a whole number portion and a fraction portion is said to be a mixed number. For example, in the following mixed number, the number 4 is the whole number and the rest is the fraction.*

$$4\frac{1}{5}$$

N

negative	*The opposite of positive. The word* negative *in mathematics usually refers to a number, such as −9, found on the negative (or left) side of zero on the number line.*

numerator	*In a fraction, this refers to the number on the top of the fraction. For example, in the following fraction, the number 3 is the numerator.*

$$\frac{3}{5}$$

O

operation	*Usually refers to the four basic arithmetic operations:* $+$, $-$, x, *and* $\div$.

P

pattern	*Pattern usually refers to something that repeats. In other words, for something to contain a pattern, there needs to be something that is repetitious.*
place value	*Usually refers to the concept that the value of each digit in a number is determined by its position in the number.*
polynomial	*A polynomial is more than one term connected with either the operation of addition or subtraction. There are many kinds of polynomials, binomials, and trinomials being the most common. An example of a polynomial is*

$$x^2 + 2x - 9.$$

positive	*The opposite of negative. The word* positive *in mathematics usually refers to a number, such as 9, found on the positive (or right) side of zero on the number line.*
probability	*Probability is the study of chance. Probability theories are used to determine how likely something, called an event, will occur or not occur.*
product	*The answer when two or more numbers are multiplied together.*

Q

quotient

The answer when a number is divided by another number.

R

reciprocal

The reciprocal of a number is found by flipping it. Flipping is defined as taking a number and switching its numerator's location with its denominator's. For example, the reciprocal of $\frac{1}{3}$ is 3

Please note also that the reciprocal of 3 is $\frac{1}{3}$.

S

sine

One of the six fundamental trigonometric functions. The sine of a right triangle is usually described as a ratio or fraction. The ratio is of the opposite side of a triangle over that of the hypotenuse.

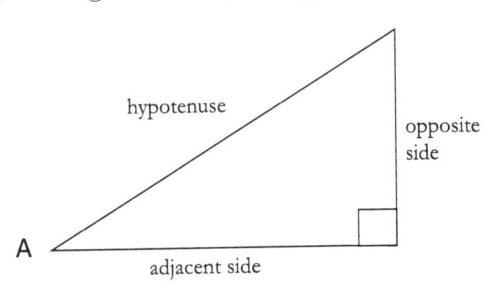

The diagram above is of a right triangle. If we were to find the sine ratio, we would put the length of the opposite side over the length of the hypotenuse.

statistics

Statistics is the study of data. Statisticians collect and analyze the data for trends. This data is usually collected and displayed in the form of graphs, charts, or tables.

sum

The answer when two or more numbers are added together.

T

tangent

One of the six fundamental trigonometric functions. The tangent of a right triangle is usually described as a ratio or fraction. The ratio is of the opposite side of a triangle over that of the adjacent side.

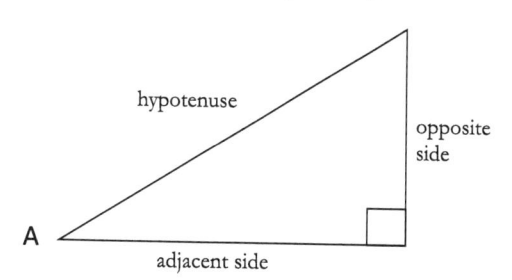

The diagram above is of a right triangle. If we were to find the tangent ratio, we would put the length of the opposite side over the length of the adjacent side.

transcendental number

A number that is not algebraic, or in other words, it is not a root of a nonconstant polynomial equation with rational coefficients. Basically, it is a number that is not an answer to a polynomial equation. Examples of transcendental numbers are π and e.

U

undefined

This term is used to define something that happens that should not or cannot happen. Seems ludicrous and yet makes sense at the same time. In other words, when something happens that interferes with a concrete rule in mathematics, we state whatever made it happen as undefined. The example in this book of something undefined is when trying to divide by zero. Fun things happen at points where graphs, polynomials, or equations are undefined.

V

variable

In the simplest of definitions, it is the unknown of an equation or polynomial. The most common variable in algebra is the letter x. However, any letter or symbol can be used as a variable as long as it is defined as such.

BIBLIOGRAPHY

The following resources were used as reference for developing the ideas and explanations found in this book. Each of these resources is of benefit to all and would be a great addition to the libraries of those fascinated and curious about the world of mathematics.

Bacon, Roger, and Robert Belle Burke. *The Opus Majus*. Philadelphia: University of Pennsylvania, 1928.

Barnes-Svarney, Patricia L., and Thomas E. Svarney. *The Handy Math Answer Book*. Detroit: Visible Ink, 2006.

Bell, Eric Temple. *Men of Mathematics*. New York: Simon and Schuster, 1937.

Bellos, Alex. *Here's Looking at Euclid: A Surprising Excursion through the Astonishing World of Math*. New York: Free, 2010.

Buchan, Jamie. *Easy as Pi: The Countless Ways We Use Numbers Every Day*. Pleasantville, NY: Reader's Digest Association, 2009.

Caine, Renate Nummela, and Geoffrey Caine. *Making Connections: Teaching and the Human Brain*. Wheaton, MD: ASCD, 1991.

Cajori, Florian. *A History of Mathematical Notations*. Chicago: Open Court Pub., 1928.

Chabert, Jean-Luc, and E. Barbin. *A History of Algorithms: From the Pebble to the Microchip*. Berlin: Springer, 1999.

Clawson, Calvin C. *Mathematical Mysteries: The Beauty and Magic of Numbers*. New York: Basic Books, 1996.

Common Core State Standards Initiative. May 19, 2011. http://www.corestandards.org/.

Enzensberger, Hans Magnus, and Rotraut Susanne Berner. *The Number Devil: A Mathematical Adventure.* New York: Henry Holt, 1998.

Foundations for Success: Findings and Recommendations from the National Mathematics Advisory Panel. Washington, DC: National Mathematics Advisory Panel, 2008.

Fulton, Brad, and Randy Phillip. "Answering Your Students' Why Questions in Mathematics." *Teacher to Teacher Press.* March 29, 2011. http://tttpress.com/pdf/Answering-Your-Students-Why-Questions-in-Mathematics.pdf.

Greenes, Carole E., and Rheta N. Rubenstein. *Algebra and Algebraic Thinking in School Mathematics.* Reston: National Council of Teachers of Mathematics, 2008.

Ifrah, Georges. *From One to Zero: A Universal History of Numbers.* New York: Viking, 1985.

Maor, Eli. *E: The Story of a Number.* Princeton, NJ: Princeton UP, 1994.

Pickover, Clifford A. *A Passion for Mathematics: Numbers, Puzzles, Madness, Religion, and the Quest for Reality.* Hoboken, NJ: Wiley, 2005.

Pickover, Clifford A. *The Math Book: From Pythagoras to the 57th Dimension, 250 Milestones in the History of Mathematics.* New York, NY: Sterling Pub., 2009.

Posamentier, Alfred S., and Ingmar Lehmann. *Pi: A Biography of the World's Most Mysterious Number.* Amherst, NY: Prometheus, 2004.

Reshaping School Mathematics: A Philosophy and Framework for Curriculum. Washington, DC: Mathematical Sciences Education Board, National Research Council, 1990.

Schifter, Deborah. *What's Right About Looking at What's Wrong?* Educational Leadership Magazine, Association for Supervision and Curriculum, November 2007.

What We Know about Mathematics Teaching and Learning. Bloomington, IN: Solution Tree, 2010.

INDEX

A

absolute value, 70-72
abstractness, 15
algebra, 14-16
algebra tiles, 81
algorithm, 15
al-jabr, 14-15
al-Khowarizmi, Muhammad ibn Musa, 14-15
Hidab al-jabr wal-muqubala (The Book of Restoration and Balancing), 14
On Calculation with Hindu Numerals (Algoritmi de numero Indorum), 15
angle measure, 94, 96
angles, 19, 24, 95, 97
arc length, 94, 96-97, 99
arithmetic, 12

B

Babylonians, 95
Bacon, Roger, 12, 103
base ten blocks, 26-27

C

calculus, 20-21
Cardano, Gerolamo, 93

circle
    circumference of, 99
    degrees in, 94-95
    radians in, 96-102
Common Core State Standards, 7, 35
completing the square, 80-82
conic section, 80

D

decimals
    dividing by, 54-56
    multiplying with, 51-53
degrees, 94-98
Descartes, Rene, 93
distributive property, 78
division
    and multiplication, 31-32, 55
    as repeated subtraction, 32, 37, 42, 54-56
    and subtraction, 74-75

E

e, 89-91
Euclid, 17
    The Elements, 17
Euler, Leonhard, 90
Euler number. See e

F

factorials, 83-85
Fibonacci, 22-23
    Liber Abaci, 23
FOIL, 76-79, 103
fractions, 45
    dividing by, 5, 42-44
functions and graphs, definition of, 16

G

geometria, 17
geometry, 17-18

H

Hidab al-jabr wal-muqubala (The Book of Restoration and Balancing) (al-Khowarizmi), 14
Hindu-Arabic numerals, 23-24
hundreds block, 26

I

i, 92-93
improper fraction, 45-50
infinitesimal analysis. *See* calculus
infinity, 33
integers, multiplying with
    negative and negative, 61-62
    positive and negative, 59-60
invert and multiply, 42

L

Liber Abaci (Fibonacci), 23
logic, 12
long division, 35-41

M

máthēma, 11
mathematics
    definition of, 11-13, 103
    mysteries in, 5-6
mathmaticus, 11
Mesopotamia, 17
mixed number, 45-50
multiplication
    and addition, 74-75
    and division, 31-32, 55
    as repeated addition, 32, 57, 61
muqubalah, 14

N

$n^0 = 1$, *63-65*
$n^{-1} = 1/n$, 66-69
Napier, John, 90
numerals, writing, 22-24

O

On Calculation with Hindu Numerals (Algoritmi de numero Indorum) (al-Khowarizmi), 15
ones block, 26

P

PEMDAS, 73-75
perfect square trinomial, 80
permutation, 83-84
pi, 87-88
place value, 30
products, 57, 59, 61
Pythagorean theorem, 17

R

radian, 96-102
radius, 96-97, 99
regrouping, 25-30
roman numerals, 22-23

T

tens block, 26
trigonometric functions, 19
trigonometry, 19

U

undefined, 31

Z

zero
    dividing by, 31-34
    multiplying with, 57-58
0! = 1. *See* factorials

Made in the USA
Lexington, KY
06 February 2012